麵包鬆軟美味，滿室芳香
The smell of fresh bakery fills the air

Contents 目錄

Fruity Bread

from the Basi

獨角仙的 簡介

· 曾任職酒店達六年
· 日本鳥越製粉麵包學校短期留學
· 台灣鐵能社麵包學校短期留學
· 烘焙愛好者
· 烹飪導師
· 著作有《豈能戒甜》、《無包不歡》、《點點我心》

　　我的網名是「獨角仙」，餅友們稱呼我為「亞堅」、「仙姐」；其實我小時候已給喚作「麵包頭」，全因任職麵包廠的外公每次帶來賣剩的麵包，皆使我雀躍不已。年紀漸長，父親常買的炸咖喱牛肉包、我在酒店工作時常常購買的小法包，皆是我人生中永不忘懷的好滋味。

　　1995年，我開始被烘焙深深吸引，狂熱地上課、看書，以求涉獵各種烘焙知識；但做過各種花巧、千變萬化的糕點甜品，還是做麵包讓我最沉醉——麵包的生命力像引發我的能量，每次做新麵包，一步一步做麵種、搓麵、發酵、造型、烘焙……還未嚐到出爐麵包，也不能確定麵包是否成功。

　　記得第一次做麵包是跟着麵包書做的，當一盤小小餐包出爐，真是興奮莫名！當時沒有網絡和這麼多的烘焙店，無論學習和購買食材也非常困難，很多時碰到問題也乏人問津，得全靠自己摸索。我買來大量書籍、影帶、工具學習，甚至訂購外國書籍，對做麵包的興趣一直有增無減，加上外子是略懂西廚，他對味道和食材的觸覺，給我帶來不少靈感；然而，畢竟是孤軍作戰……

　　直至2005年，我在網上討論區認識了一群志同道合的朋友，才知道小小的香港，愛做包餅的人何其多。初時大家在那裏交流心得，互相鼓勵，後來幾乎每星期一次的聚會，萍水相逢的朋友結成莫逆，讓我的人生進入最豐盛和滿足的時刻。一些長居外國的朋友更不時來港探望，甚至帶來外國食材和工具。去年我在電視上看到瑞士阿爾卑斯山居民仍用磚窰焗麵包，和朋友商議後，一呼百應，經過八天工程，磚窰完成了！儘管仍在摸索爐溫和技巧，但它為我們一眾愛包之人帶來無限快樂，箇中樂趣也值得回味。

　　做麵包不能太隨意，也不需要太執着，要視乎季節、溫度、食材等等因素，根據自己的經驗調節，跟發酵中的麵糰溝通——說穿了，正是用心烹調之道。直接法、冷藏法、湯種、中種各種方式，配合食材、心思、創意，人人皆能在家中烘出美味麵包與親朋分享。我在本書介紹的17小時低溫冷藏中種，具發酵芬芳，成品柔軟，變化多端，而且製作簡便，很適合繁忙的香港人在家製作。這本書是我做麵包至今的小小里程碑，希望藉此與更多愛包朋友分享做麵包的樂趣。

　　謹此感謝料理達人丘桂玲小姐的引薦，讓我的第一本書得以面世，還有多位在烘焙路上遇到的良師、一直支持和幫助我的一眾「麵包莫逆」、常常充當白老鼠的家人和朋友，和互相扶持十多載的丈夫。

獨角仙

　　認識阿堅是她往黃克競報讀我的十堂餅面設計課程，期間留意到她有做餅的天分。她的學習態度認真，每做一步都講求完美，閒談間得知這是她的業餘興趣，及後在一些特別的盛會，她也曾到學校幫手做西點。與她相識差不多十年了，我出版的三本西點書她也有請纓幫忙。作為一個業餘人士，她的熱誠比一些專業人士更專業。她很喜歡鑽研所有有關飲食的製作，而且材料運用得宜，看她收藏飲食書籍的種類和數量，便知她對飲食之要求，如麵包、西點、朱古力，甚至中菜和西菜。

　　她努力研究製作之麵包，比一些麵包店更精美呢！

香港烘焙專業協會會長

　　跟獨角仙 (亞堅) 的相識與相認，是一件很戲劇性的事情！

　　認識她，也是透過她的網上日誌，知道她對於烘焙，是一個瘋狂份子，在網上看她與幾位好友一手一腳把一個石爐造好。追看她的網誌，漸漸認識到她凡事講求認真，是性情中人。對她產生了一份敬佩之心外，亦很想結識現實中的她。

　　去年底，出席亞堅每年一度的blogger聚會，認識了她。她很隨和，不拘小節。聚會結束後，歷歷在目。沒想到，當晚收到她的留言，提到了中學時代的一些往事。那一刻我簡直不敢相信，原來大家在初中時是同班同學。那種感覺，難以言喻，一直在網絡上互相認識，然而卻原來大家一早已認識的。

　　我喜愛她對烘焙那份熱誠，專業而認真，愛分享。她並非一位專職導師，可是桃李滿門，那就是她不計較和愛分享的寫照了。

　　知道她要出版個人烘焙書籍，很替她高興！每次看她帶領一群blogger一起做出香噴噴的麵包，從焗爐裏端出來，只要嘗過一口，那份滋味，非筆墨可以形容的。

　　很想她透過書本，將那份爽朗豪情，跟讀者們一起分享她的成品，讓每位讀者都可以感受她那份從烘焙而來的喜悅！

　　我熱熾期待亞堅為我們造的書，一起見證和分享她的努力！

美味與快樂追尋

　　諾貝爾物理學得獎者丁肇中教授說過：「滿足好奇心是人生的最大樂事」。我認識獨角仙快十年了，雖然她只是烘焙業餘愛好者，但對烘焙的熱誠、追求、執着、渴求、嚴謹，有着丁肇中教授做物理實驗一樣的身影，我也相信獨角仙追尋美味的同時，也滿足了無比的好奇心，得到了無比的人生樂趣。

從您身上我見到您對烘焙作業的熱誠

　　藍色大門是您的烘焙天地，經常與工友埋首試作、研究配方、分享心得，您的熱誠感染到更多人對烘焙產生興趣，大門背後充斥着拍麵糰聲、計時器響鬧聲、歡笑聲、尖叫聲、起哄聲、相機快門聲，無論是多番成功或無數失敗，都是令人羨慕的時光。

從您身上我見到您對烘焙技術的追求

　　除了不放過參加任何香港星級餅師的烹飪班外，舉凡有外國名師來港示範烘焙技術，您會第一時間報名，馬上請假；出席當日您總會早到，您總是坐在最前排坐位，您會帶來名師著作給他們簽名，您會抓緊機會發問。

從您身上我見到您對烘焙用料的執着

　　日本麵粉、發酵牛油、上白糖、三溫糖、天然海鹽、新鮮酵母、陳醋、橄欖油、番茄乾、法國合桃、以色列棗子、宇治抹茶；水仙花麵粉、三象粘米粉、龍皇杏、蓮茸及糖水桂花等等，不論中式西式，您都會選用最好的材料，因您明白只有最好的材料，才可做出最好的產品！

從您身上我見到您對烘焙知識的渴求

　　您收藏中外烘焙書籍之多，重得連書架也快要塌下來了。還有散佈在家中角落的無數烹飪期刊和烘焙雜誌，見證了您於烘焙知識方面的增長，所以才會追求十七小時麵包的口感，才會追求頂角吐司的成功喜悅。

從您身上我見到您對烘焙工具的嚴謹

　　無論是剷麵糰的膠刮板，或是一個入爐用的麵包籃，您都有所要求，力求專業。為了追求更佳的烘焙效果，在各方好友、「瘋婦傻婆」的協助下，竟然在家園內自己動手做出「型爐」，一磚一石，毫不馬虎，開心滿足之餘，更為自己的烘焙技術設下擂台，迎接更大的挑戰。

　　一個行業的大師，就是要幫助新人成為大師。您的新書正正讓更多烘焙人士得到啟發，享受烘焙所帶來的喜悅和樂趣，向着大師的路邁進！

香港烘焙協會副會長

清晰記得第一次步入堅家,明明是她領我進屋,我仍以為自己摸錯了門—— 一雙業餘的夫妻,因為喜歡烹飪而把大半個家變成了廚房。然而,對比堅十多年來學習中西點的努力不懈與熱情,這廚房的魅力立刻相顧失色。

堅有着火車頭般的衝勁、能量、帶動力,腦裏滿載一卡一卡食材、中西點的知識,無論做甚麼都勇往直前,不成功不死心。初認識她,她帶我到家中糧倉,從碩大的麵粉袋中抓起一把麵粉湊到鼻端,說:快來聞聞——多香啊!呆頭呆腦的我當時不大感到甚麼麵粉香,卻被她面上那單純的快樂與生命力怔住了。後來,一群朋友常常聚集在「藍色大門」(我給她家起的「暱稱」),每次大家為美味的出爐麵包、甜點和大師(阿堅丈夫)的料理讚嘆時,總能看到堅和大師臉上泛起這興奮的神情。

堅這火車頭不斷策勵大家在烘焙興趣上更進一步,並毫不吝嗇地傾出所知。得知她正努力編撰一本深入淺出的麵包書,我更能略盡綿力,興奮之情難以言傳。願讀者憑着本書,發掘更多做麵包的樂趣,與家人分享健康和喜悅,也期待日後堅腦裏那些一卡一卡的西點、中點食譜和知識,陸續編撰出版。

書籍面世適逢我的孩子出生,正像快樂和希望,紛至沓來。

Cass

我這隻深居井底的小蛙,兩年前,跳出井走到「藍色大門」叩門鈴。藍色大門是烘焙好友的練功房。女主人阿堅,有一位擅於下廚的丈夫,一應俱備的廚房,一班熱心鑽研美食的女工,她把做麵包和糕點的熱誠,感染到每一位女工。最令人難忘的是:四出採購、集思研制食譜、各人親手炮製、最後品嘗美食和分享成果。整個製作過程嚴謹得來又帶點隨意,更因為要選用最合心意的烤爐,於是,總動員建造「第一型爐」。每次聚會後,我都會帶着意猶未盡的「下廚樂」心情回家。

麵包是香港人主食之一,能為家人朋友送上親手製作新鮮出爐麵包,看着他們在進食時流露的笑容,那份滿足感令人難以忘懷。要做出美味又健康的麵包,比糕點製作更難,單看一兩本食譜,很不容易才掌握到箇中的要訣。阿堅有見及此,在書中加入不少篇幅,希望新手們能更快學會麵包製作的基本功,享受到自家製作和分享成果的喜悅。

給你一句說話:阿堅你係得架!

WaWa

一切由基本開始
Start from the basics

　　首先深入淺出地讓大家認識製作麵包的原材料對成品質素的影響；而烘焙百分比看似複雜，但實際是很簡單的，只要你能掌握計算方法，便可以按照所需成品份量，計算出材料用量；至於本書的靈魂17小時低溫冷藏中種，它令麵包有嚼勁、濕潤，帶有發酵的芬芳，可存放較長時間而不易變硬，保濕良好。

　　有了基本的認識，是時候動手了。跟着麵包製作的12個基本步驟，來製作最基本的白吐司，當你成功了，就可挑戰其他麵包。

Before starting, you should familiarize yourself with how the quality of bread ingredients affects the end-product. After that, try to practice with the baker's percentage. Although it seems complicated, it's actually much simpler than you think. As long as you have no problem converting the figures according to baker's percentage, you can easily calculate the amount you need for each ingredient. Finally, the most crucial basic skill you need is the "17-hour low temperature" pre-ferment dough. It is used in all bread recipes throughout this book and it makes the bread chewy, moist, yeasty and have a longer shelf life.

After you feel comfortable with the basics, it's time to get hands on. Follow these 12 basic steps in bread making and you'd end up with your very first loaf of White Toast. When you're ready for more challenges, feel free to make other kinds.

麵包的基本材料
Basic bread ingredients

麵包的基本材料有麵粉、酵母、水、鹽、糖、蛋、油脂和奶,雖然材料少,但麵包的味道、質感和造型,視乎你的魔術手怎樣處理它們。

The basic ingredients of bread include flour, yeasts, water, salt, sugar, eggs, grease and milk. Although it isn't exactly rocket science, bread can be versatile in terms of texture, taste and shapes, depending on the proportion of different ingredients and the way you handle them.

麵粉 Flour

做麵包主要是用高筋粉(或稱筋粉),它由小麥磨製而成。麵粉所含的蛋白質經水混合後產生麩質,經搓揉後成為麵筋,形成支撐整個麵包的主要結構。麵粉以所含蛋白質的多寡來區分,高筋粉含11.5%-13.5%蛋白質,一般麵包製作主要使用高筋粉;中筋粉含蛋白質8.5%-11.5%,通常用於製作批皮、麵包;低筋粉含少於8.5%蛋白質,用於製作蛋糕、餅乾。

高筋粉顏色較黃,質感輕爽和有光澤,用手抓不易黏成一團。因各品牌不同而吸水量有差別,所以在轉換慣用的麵粉時務必把配方中的水分預留一些,以作調整。

本書使用中種法,麵粉(包括裸麥粉、全麥粉等)在本書各食譜中,中種和主麵糰的麵粉加起來為100%。其他材料的用量以麵粉重量為基數來推算。(請參考第16頁的「烘焙百份比」)

As its name suggests, bread flour (or high-gluten flour) is commonly used in bread. It is ground wheat with high protein content that turns into gluten after being mixed with water. After kneaded repeatedly, the dough becomes resilient as the gluten supports its structure. Flour is classified according to their protein content. Bread flour has 11.5% to 13.5% of protein and is used in bread making; all-purpose flour is 8.5% - 11.5% protein and is used in pie crust and certain bread; cake flour contains less than 8.5% protein and is used in cakes or cookies.

Bread flour is yellowish in colour, light in texture and shiny. It's not easy to squeeze it into a lump. Depending on the brand, different bread flour may absorb water differently. Thus, if you're using a certain brand of flour for the first time, make sure you don't add all the water at one time. Save a little to allow adjustment. The recipes in this cookbook use pre-ferment dough as a base. The total weight of all flours (including rye flour and wholewheat flour) add up to 100%. The weights of other ingredients are calculated as a percentage of the total weight of flours. (Please read "baker's percentage" on p. 16).

酵母 Baker's Yeasts

酵母由微生物培養出來，它為麵包賦予生命。酵母得到水和養份，在適合溫度下便能繁殖。酵母菌繁殖時產生二氧化碳，使麵糰膨脹。市面上的酵母可分鮮酵母、乾酵母和速效酵母三種。

鮮酵母含有大量水分，必須存放在低溫的環境。在0-5℃雪櫃可存放一個月，如存放在冷藏櫃雖然可以保存較久，但酵母的活力會大為減弱，用量亦要相應增加。鮮酵母濕度約有70%，發酵速度也較快，可直接加於麵粉一起攪拌，使用方便。本書建議使用鮮酵母，因它的風味最佳。如酵母顏色有異或味道轉變，表示酵母已經變壞，不能使用。

乾酵母是將鮮酵母壓榨成細小圓粒狀，再將其風乾，令它成休眠狀態，不易變質。乾酵母使用時只須加入少量溫水和糖浸泡約10-15分鐘令它軟化和恢復活力，才可加入其他材料中攪拌。它的特性是發酵耐力持久，但由於使用不方便，故使用不廣。乾酵母的用量是鮮酵母的一半，酵母味道濃郁。

速效酵母的製作和乾酵母相同，呈細小短條粉粒狀，可直接加入材料中攪拌，使用方便。如製作少量麵包，建議購買超級市場個別包裝的速效酵母。因酵母一經開封，便會慢慢失去活力，這情況往往在你很用心地搓完麵糰後才會察覺。速效酵母用量是鮮酵母的三份之一。

不論哪種酵母，用剩的都應存放在雪櫃。

Yeasts are microorganisms in the fungus kingdom. They are used in bread as a leavening agent to make it fluffy. When yeasts are given water and nutrients, they quickly reproduce under the right temperature. Carbon dioxide is produced in due course which would expand when the dough is heated up in the oven. There are three kinds of yeasts in the market, namely fresh yeasts, dried yeasts and instant yeasts.

Fresh yeasts – have high water content and should be kept at a low temperature. They last in the fridge for 1 month at 0°C to 5°C. Although they last much longer in the freezer, the freezing step tends make them less active and you have to use a larger quantity to make up for that. Fresh yeasts contain about 70% water and they tend to prove quickly. They can be added straight into the flour before mixing and kneading. I suggest using fresh yeasts in all recipes included in this book because of its unique taste. If the yeasts look or smell different, they might have gone stale and you should discard them.

Dried yeasts – Fresh yeasts are pressed into small balls and dried. The yeasts would then lie dormant and have a much longer shelf life. Before using dried yeasts in dough, you should soak them in warm water with sugar added for 10 to 15 minutes, to wake them up. Dried yeasts have a lasting leavening power but it's not commonly used by professional bakers because of the additional "waking" step. If you use dried yeasts for the recipes in this book, use only half the amounts listed for fresh yeasts. Dried yeasts also make the bread taste strongly yeasty.

Instant yeasts – They are made in similar ways like the dried yeasts and they are in short strip or pellet form. You don't have to wake them up before using and you may put them directly into the dry ingredients before mixing. If you're making a small quantity of bread, it's advisable to get instant yeasts in individual small packets from supermarket. Once a packet of yeasts is open, they slowly lose their leavening power, and it's hardly noticeable until you've spent all the time kneading the dough. Thus, try your best to use up the whole packet and chute any leftover. In case you use instant yeasts for the recipes in this book, use only 1/3 the amounts listed.

No matter which yeasts you use, always keep the leftover in the fridge.

蛋 Eggs

蛋可增添麵包的香味和提高營養價值，令麵包增加彈性、光澤、質地鬆軟。蛋脂肪中的卵磷脂，為水和油的中介，可當作天然乳化劑，並幫助成品保持濕潤。

Eggs make the bread more fragrant and nutritious. Eggs also make the bread fluffy and glossy. The lecithin in egg fat is a medium for both water and oil. It is also a natural emulsifier that keeps the baked goods moist.

水 Water

它是麵粉、酵母以外製作麵包不可缺少的材料。麵粉與水搓揉後能產生筋性，酵母也需要水分才能發酵。本書用水量很高，約在60%-69%，初學者或用手搓麵糰者可將水分減少5%-8%。如改用蔬果汁或含有水分的材料，便要相應減少水量，以免麵糰過濕。

An indispensable ingredient in bread making apart from flour and yeasts, water gives flour elasticity after kneading. Water is also the essential medium for yeast fermentation. The recipes in this book call for a high water content in dough, which is usually around 60% to 69%. Beginners and those who knead their dough with their hands may reduce the water content by 5% to 8%. In case you use vegetable juice or other ingredients with high water content, you should further reduce the amount of water. Otherwise, your dough will be too wet.

Basic
bread
ingredients

奶類 Milk products

奶可以增加麵包的風味、營養價值和色澤，並增加成品的柔韌性。奶類製品如淡奶、煉奶、鮮奶和奶粉也普遍使用於麵包製作中。因鮮奶保鮮期有限，所以奶粉是最好的代用品。

Milk gives bread a special taste; enhances its nutritional value and colour; while making it more chewy and spongy. Milk products like evaporated milk, condensed milk, regular milk or milk powder are commonly used in bread making. As fresh milk has limited shelf life, milk powder is considered the best substitute.

油脂 Grease

油脂可改善麵包的品質，在麵糰發酵時發揮潤滑作用，促進麵包膨脹，令麵包鬆軟，並可延長成品保存時間，增加香氣和營養價值。一般麵包的油脂使用量約在6%-12%。油脂過多會令麵包內部組織粗糙，並延緩發酵速度。本書使用的油脂是無鹽牛油和橄欖油。

Grease enhances the texture of bread and lubricates the dough in the proving process. Grease also helps raise the dough; makes the bread fluffy; extends the shelf life of baked goods; and enhances their fragrance and nutritional values. Most bread has a fat content around 6% to 12%. Adding too much fat would make the bread tough and retard the proving activity. The grease used in the recipes here is either unsalted butter or olive oil.

鹽 Salt

鹽是麵包製作的一個重要元素，它不單可增加麵包的香味，還能在麵包的發酵過程中強化麵糰中的麩質，改善麵糰的柔韌性和彈性，增加麵糰的膨脹力。鹽能抑制酵母，控制麵糰發酵比率；沒加入鹽的麵糰發酵較不穩定，容易發酵過度。一般麵包的鹽量在0.8%-2.5%之間，過多會過鹹和過度抑制發酵。

本書所有食譜均使用海鹽，海鹽令麵包的風味更佳。

Salt is another essential ingredient in bread. Not only does it enhance the fragrance of the bread, it also strengthens the gluten structure in the proving process to give the dough softness, resilience and leavening power. Salt suppresses the action of yeasts and is used to control the leavening ratio. Dough without salt tends to be less stable and may be over-proved. Most bread has a salt content of 0.8% to 2.5%. Adding too much salt would make the bread too salty and insufficiently raised.

All recipes included in this book are using sea salt because of its unique taste.

糖 Sugar

糖是提供酵母養分的主要來源。除了增加風味，使麵包柔軟，增進色澤，更能保持成品濕潤，加強防腐。各種糖類甜度和風味各有不同，只要掌握它們的性質，就可自由搭配出不同的口味。

Sugar provides the nutrients for yeast fermentation. Besides adding taste, sugar also makes the bread soft and gives a nice colour. Sugar also keeps the bread moist and acts as a natural preservative. Different sugars are different in sweetness and taste. By using the right kind and amount of sugar, you may then create your very own signature taste.

甚麼是17小時中種冷藏法？

What is 17-hour low-temperature pre-ferment dough?

中種法是麵包發酵方式的一種，將食譜的份量分前後兩段時間攪拌，前段通常是將60%至85%的麵粉、部分的酵母、鹽或糖和水先攪拌，再經3至4小時的發酵成為中種，接着加入主麵糰的材料（15%-40%麵粉及其餘材料）攪拌，經短時間的延長發酵，成為麵糰。

有很多不同的中種法，本書介紹的17小時麵包屬於中種法中的低溫冷藏法；麵糰的70%份量經搓揉後存放在0-5℃雪櫃內，經17小時冷藏發酵成中種麵糰，主麵糰的發酵時間相應地減少，最後發酵後勁凌厲，但烤焗的時間比較長。

Pre-ferment dough is a way of proving in the bread making process. Divide all ingredients in a recipe into two parts. The first part usually includes 60% to 85% of the flour; part of the yeasts; salt or sugar; and some water. Mix this first part well and let it prove for 3 to 4 hours. This is called the pre-ferment dough. Then the rest of the ingredients are added (15% to 40% of the total amount of flour and other ingredients). Mix well and let it prove for the second time very briefly. The dough is then ready for baking.

There are many different methods to produce pre-ferment dough. The method used in all bread recipes in this book is called the "17-hour low-temperature" method. 70% of the dough is kneaded and stored in a fridge at 0-5°C for 17 hours. As most of the dough has been proved once under low temperature, the final proving time is greatly reduced and the final proving gives a major raise in the dough. Yet, you need to bake the bread for a longer period.

這種發酵方法的成品，特質是口感有嚼勁、濕潤，帶有充分發酵的芬芳，可存放較長時間而不易變硬，保濕良好。而且口味隨意配搭，是一種千變萬化的包種，只要加少許創意，你也能輕易做到在一般麵包店買不到的可口麵包。

別以為做這冷藏中種法麵包很費勁，其實它是最方便不過的，很適合上班一族。我們預先一晚做好中種麵糰放雪櫃內發酵，第二天下班後完成其餘工序，而且可先做好中種麵糰，再慢慢考慮麵包的味道。如果第二天不巧沒時間也不打緊，麵糰可以雪藏達72小時。所以學懂這麵包的做法，只要一星期動一天功夫，就可以天天有新鮮兼適合自己口味的麵包了。

以下是製作冷藏中種法的步驟圖：

1. 將酵母溶於水中，再加入筋粉、海鹽、脫脂奶粉搓至柔滑成麵糰(看圖1-5)，用保鮮紙包裹麵糰。

2. 放入雪櫃發酵17小時(看圖6-7)。

註：

- 圖6的(a)是已發酵的麵糰，而(b)是剛搓至柔滑的麵糰

- 圖7是發酵麵糰的橫切面

The bread made with pre-ferment dough has a moist and chewy texture and a yeasty fragrance. It also lasts long without turning hard or getting dry. In terms of taste, you can easily match it with almost any baking ingredients. As long as you are creative enough, you'd end up with your signature bread not available elsewhere that tastes better than anything you can get from a store.

Don't be tempted to think this pre-ferment dough is a drag. It's actually extremely easy and convenient to make and it especially works well with the busy schedule of working urbanites. Regardless of the taste of the bread or ingredients you intend to use, you'd still need the same pre-ferment. So it doesn't hurt to make it the night in advance and leave it in the fridge for low-temperature proving. You have all day to think about what flavour you want the bread to be. On the next day when you come home from work, just finish off with the quick second proving and bake it. And voila, freshly baked bread is done. Even if you don't have time the next day, the pre-ferment dough lasts well in the fridge for 72 hours. In other words, you can make a whole bunch of pre-ferment dough in one day of the week, and you'd have fresh bread baked to your own taste for the rest of the week. Isn't that great?

Here are illustrations on the method of making pre-ferment dough:

1. Dissolve the yeast in water. Add bread flour, sea salt, skim milk powder and knead until soft. Cover the dough in cling wrap (see pictures 1-5).

2. Refrigerate to let it prove for 17 hours (see pictures 6-7).

Note:

- Picture 6a shows the dough after proving. Picture 6b is the smooth dough before proving.

- Picture 7 is a cross-section of the proved dough.

烘焙百分比
Baker's percentage

製作麵包，可大可小——説的是材料與成品份量。

同一種麵包，只要按着「烘焙百分比」，便可以按照所需成品份量，計算出材料用量，即是説同一款麵包，無論麵包店製作一天供應量，還是家庭製作一小盤，材料的比例也是一樣的。

烘焙百分比以麵粉重量為基數，即是不管麵粉重多少，我們也把它設定為100%，其他材料的重量按照麵粉的百分之幾來計算。

As long as you're making the same bread, no matter you bake one loaf or one thousand loaves, you may easily calculate the amount of each ingredient with baker's percentage. To determine the baker's percentage, the weight of flour is always set at 100%. All ingredients are measured by their weight in relation to that of flour. The total must add up to over 100%.

例子一：按照食譜上的材料重量，計算烘焙百分比

（材料重量/麵粉重量）x 100 = 烘焙百份比

Example 1: calculate the baker's percentage according to the amount listed in a recipe

Baker's percentage = (weight of a certain ingredient / weight of flour) X 100

材料 ingredients	重量 weight	烘焙百分比 baker's percentage
麵粉 flour	1000 g	100%
糖 sugar	30 g	(30/1000 x 100) = 3%
雞蛋 egg	200 g	(200/1000 x 100) = 20%
海鹽 sea salt	10 g	(10/1000 x 100) = 1%
酵母 yeast	30 g	(30/1000 x 100) = 3%

例子二：按照烘焙百分比，計算各項材料的重量

麵粉重量 x 烘焙百分比 = 材料重量

Example 2: calculate the weight of individual ingredients according to baker's percentage

Flour weight X baker's percentage = weight of individual ingredient

材料 ingredients	烘焙百分比 baker's percentage	重量 weight
麵粉 flour	100%	1000 g
糖 sugar	3%	1000 g x 3% = 30 g
雞蛋 egg	20%	1000 g x 20% = 200 g
海鹽 sea salt	1%	1000 g x 1% = 10 g
酵母 yeast	3%	1000 g x 3% = 30 g

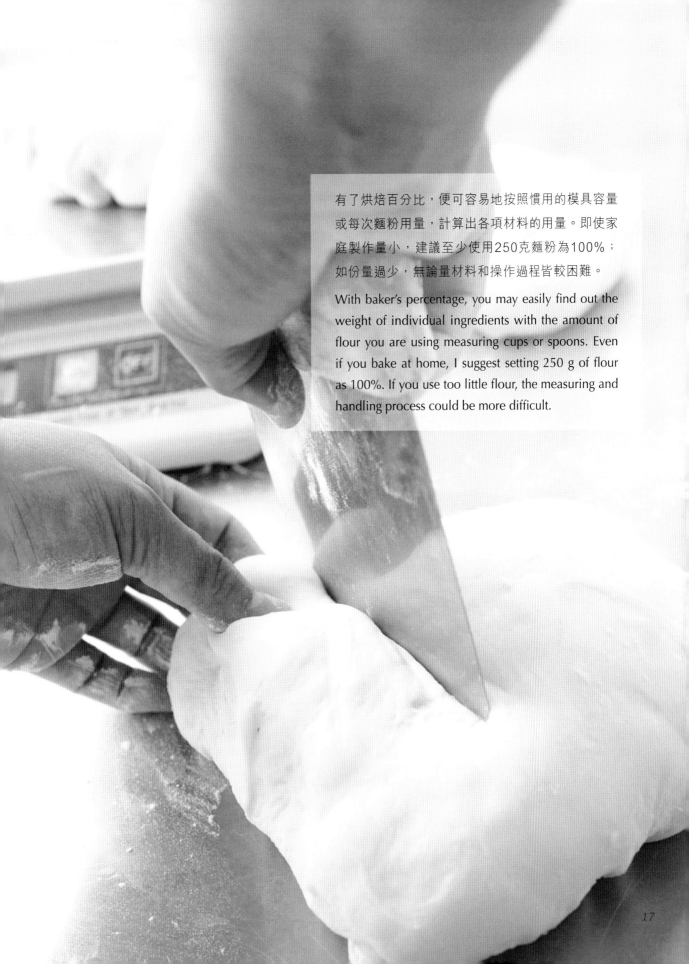

有了烘焙百分比，便可容易地按照慣用的模具容量或每次麵粉用量，計算出各項材料的用量。即使家庭製作量小，建議至少使用250克麵粉為100%；如份量過少，無論量材料和操作過程皆較困難。

With baker's percentage, you may easily find out the weight of individual ingredients with the amount of flour you are using measuring cups or spoons. Even if you bake at home, I suggest setting 250 g of flour as 100%. If you use too little flour, the measuring and handling process could be more difficult.

麵包製作的12個基本步驟
The 12 basic steps in bread making

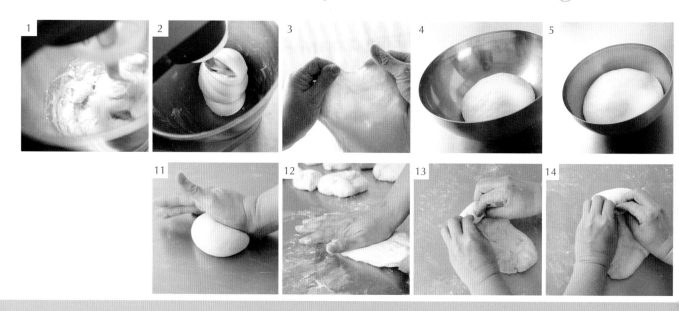

1. **計算並量度材料**

   計算好所需材料份量，使用精確的儀器，秤各種材料。

2. **混合**

   把麵粉、水、酵母、糖、海鹽等等材料混合，搓揉至有筋性，直至麵糰表面光
   滑、有彈性和可拉出薄膜。(看圖1-3)

3. **發酵**

   將麵糰放在大碗內，置於室溫下（約23℃-30℃，冬天用暖水保溫）進行發酵。
   當發酵至約兩倍大，可測試麵糰。食指沾少許麵粉，慢慢戳入麵糰中，如麵糰發
   酵適當，指孔不會收縮；如指孔迅速回縮，即發酵不足；若戳下時整個麵糰洩
   氣、收縮，便是發酵過度了。(看圖4-5)

4. **排氣**

   用拳頭擊出麵糰內之空氣，取出麵糰，把麵糰邊緣摺向中央。此動作可鬆弛麩
   質，重新分佈酵母和其他材料，並平衡麵糰溫度，令麵糰發展得更好。(看圖6)

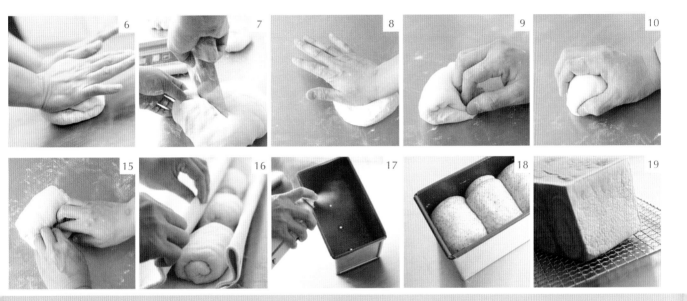

1. **Calculating and measuring the ingredients**

   Find out how much of each ingredient you need. Measure accurately with a scale.

2. **Mixing**

   Mix the flour, water, yeasts, sugar and sea salt etc together. Knead the dough until it feels elastic. It should be smooth on the outside and stretchable into a thin sheet without breaking. (see pictures 1-3)

3. **Proving (or fermentation)**

   Put the dough into a large bowl under room temperature (i.e. 23°C to 30°C; use warm water in winter). Wait until the dough doubles its size and you may test the dough. Dip your index finger into the flour. Slowly poke the dough with your index finger. If the dough has been proved properly, the indentation made with your finger should stay the same size. If the indentation collapses and flattens after you pull your finger out, it is under-proved. On the other hand, if the whole dough collapses and shrinks after you pull your finger out, the dough is over-proved. (see pictures 4-5)

4. **Driving the air out**

   Flatten the dough with your fists to drive the air out. Take the dough out of the bowl and fold the edged towards the centre. This helps relax the gluten in the dough; redistributes the yeasts among other ingredients; and evens out the temperature throughout the dough. (see picture 6)

5. 分割

   用刮刀和磅平均地分割和秤麵糰。動作要快，否則各個小麵糰會發酵不平均。(看圖7)

6. 滾圓或輕捲成條狀

   分割後將麵糰滾成圓形或輕捲成條狀，伸展麵糰至外表光滑，使完成品更有吸引力更有光澤。以下是滾圓和輕捲成條狀的圖解：

   滾圓：(看圖8-11)

   輕捲成條狀：(看圖12-15)

7. 延伸發酵

   麵糰滾圓或輕捲成條狀後，放在桌面或烤盤上，蓋上保鮮紙保濕，作伸延發酵，大約15-20分鐘，讓麵糰鬆弛，易於造型。(看圖16)

8. 造型

   如用模具，要先塗油防黏，方把麵糰放入。留意麵糰收口朝下，使用藤製麵包籃則朝上。(看圖17)

9. 最後發酵

   把麵糰放在溫暖地方發酵至適當體積。(看圖18)

10. 烤焗

    麵糰最後發酵完成，可作裝飾（如塗蛋水、灑麵粉、裹芝麻、荊上花紋等等），然後放入已預熱焗爐烘烤。每個焗爐的火力和操作都不同，所以要熟悉自己的焗爐運作以作調校。

11. 冷卻

    如使用模具烘烤，出爐後要立即脫模，並放在網架上冷卻，以免水蒸氣弄濕成品。出爐麵包內部充滿水氣，應待麵包冷卻或只有些微餘溫時方切開或食用，否則麵包組織會黏在一起，變得糊口。(看圖19)

12. 儲存

    麵包由出爐一刻開始老化。家庭製作的麵包不加入添加劑，故一般比市售麵包容易變得乾硬；但本書食譜加入中種，能有效延長保鮮期，經二至三日亦綿軟可口。若把麵包貯存於冷凍庫，可保存多月，食用前取出置於室溫解凍，用微波爐或焗爐加熱，即可回復新鮮可口。

5. Dividing

   Use a scraper or dough cutter to divide the dough. Weigh each piece of dough with a scale to make sure they are of equal weight. You should do this step quickly. Otherwise, the pieces of dough may prove differently. (see picture 7)

6. Rounding or hand squaring

   Round or hand square each piece of dough after dividing them. Each piece should look smooth on the surface for the end-product to have the best presentation. Please refer to the pictures 8 to 11 for rounding; and pictures 12 to 15 for hand squaring.

7. Extended proving

   After rounding or hand squaring, put the pieces of dough on a countertop or in a baking tray. Cover with a cling wrap. Leave the dough to prove again for 15 to 20 minutes. This step also helps relax the dough and make it easy to shape into desired form. (see picture 16)

8. Styling

   If you use a mould, make sure you grease it for easy unmoulding. Then put the dough in with the seam side facing down. Yet, if you use a proving basket, you should put the seam side up. (see picture 17)

9. Final proving

   Leave the dough in a warm spot until it expands to desired volume. (see picture 18)

10. Baking

    Decorate the dough if necessary (such as brushing on egg wash, sprinkling flour or sesames, or scoring the dough crust etc). Put it into a preheated oven for baking. As the heat and temperature of each oven vary, you should experiment with your own oven and make adjustment to the baking time and temperature.

11. Cooling

    If you bake the bread in a mould, you should unmould it immediately after it's done. Cool it on a wire rack to prevent the crust from picking up the condensation. The inside of freshly baked bread is filled with steam and moisture. That's why you should let it cool off completely or at least until you can barely feel the heat before you slice or eat it. Otherwise, the interior texture will be tangled together and you end up with a sticky mess. (see picture 19)

12. Storage

    The bread starts to age since the moment it comes out of the oven. In a home kitchen, I do not suggest adding additives to bread. Thus, the bread you make at home is more likely to get dry and hard sooner than those you get from a store. I intentionally use the pre-ferment dough in all bread recipes here because it extends the shelf life at room temperature to 2 to 3 days without getting dry and hard. If you freeze the bread in a freezer, it actually lasts for months. Just leave it to thaw under room temperature before serving. Or reheat it in a microwave or conventional oven. It'd taste as good as it's fresh.

白吐司
White Toast

材料 ingredients

中種 Pre-ferment Dough	百分比%	克 gram
筋粉 bread flour	70%	480
水 water	40%	274
鮮酵母 fresh yeast	2%	14
海鹽 sea salt	1%	7
脫脂奶粉 skim milk powder	2%	14

酵母溶於水中，再加入筋粉、海鹽、脫脂奶粉搓至柔滑，用保鮮紙包裹麵糰，放入雪櫃發酵17小時。

Dissolve the yeast in water. Add bread flour, sea salt, skim milk powder and knead until soft. Cover the dough in cling wrap. Refrigerate to let it prove for 17 hours.

麵糰 Dough	百分比%	克 gram
筋粉 bread flour	30%	206
海鹽 sea salt	1%	7
砂糖 sugar	10%	69
水 water	29%	199
鮮酵母 fresh yeast	0.5%	3
無鹽牛油 unsalted butter	7%	48

預備 preparation

- 將中種切成小塊
- Cut the pre-ferment dough into small pieces.

1

2

做法
method

混合： 將麵糰所有材料(牛油除外)混合，搓揉，逐少加入中種搓至柔滑，加入牛油再搓至可伸延薄膜。

發酵： 麵糰放入大碗內，蓋上保鮮紙進行第一次發酵，約25-30分鐘。

分割： 將麵糰分成六等份，排氣，按扁，輕捲成條狀，靜置20分鐘讓麵糰鬆弛。

造型： 將麵糰排氣，再擀薄成長方形，兩邊向內摺起，再擀薄，寬度與吐司模相若。捲起後放入吐司模，不要把蓋全蓋上，用保鮮紙蓋好(看圖1-8)。

最後發酵： 麵糰發酵約45分鐘，發起至八成滿，把蓋蓋上，放入已預熱170-180℃的焗爐內，焗約30-35分鐘至金黃。

1. Knead all ingredients together (except the butter). Add pre-ferment dough piece by piece and knead after each addition until smooth. Add butter and continue to knead until stretchable consistency.

2. Put the dough into a big bowl. Cover with cling wrap and let it prove for about 25-30 minutes.

3. Divide the dough into six small equal portions. Flatten them with your hand to drive the air out. Hand square them. Set aside to rest for about 20 minutes.

4. Flatten the dough with your hands to drive the air out. Roll the dough out into a rectangle. Fold both sides towards the centre and roll again. Roll it out again until its width is similar to that of the loaf tin. Place it into the loaf tin, but don't close the lid tightly. Cover with cling wrap (see pictures 1-8).

5. Lastly leave it to prove for about 45 more minutes or until the dough has risen up to 80% of the depth of the loaf tin. Close the lid. Bake in a pre-heated oven at 170-180°C for about 30-35 minutes.

3

4

5

6

7

8

貼士
tips

- 450克方形吐司模需約650至660克麵糰才可頂角,可以將一至數個麵糰放入一個吐司模內,本份量可製作兩個450克吐司模的吐司。

- 不是每個份量的吐司都能做到頂角,要看最後的發酵情況。

- To fill a 450 g loaf tin all the way to all top corners, you need 650 to 660 g of dough. You may fit one to several pieces of dough into a loaf tin. The quantity listed in this recipe is enough to fill two 450 g loaf tins.

- Whether the baked loaf fills the top corners of the tin depends on the proving conditions and that may not happen every time.

蘋果肉桂麵包
Cinnamon Apple Buns

材料 *ingredients*

中種 Pre-ferment Dough	百分比%	克 gram
筋粉 bread flour	70%	449
水 water	40%	256
鮮酵母 fresh yeast	2%	13
海鹽 sea salt	1%	6
脫脂奶粉 skim milk powder	2%	13

酵母溶於水中，再加入筋粉、海鹽、脫脂奶粉搓至柔滑，用保鮮紙包裹麵糰，放入雪櫃發酵17小時。

Dissolve the yeast in water. Add bread flour, sea salt, skim milk powder and knead until soft. Cover the dough in cling wrap. Refrigerate to let it prove for 17 hours.

麵糰 Dough	百分比%	克 gram
筋粉 bread flour	30%	192
脫脂奶粉 skim milk powder	3%	19
海鹽 sea salt	1%	6
砂糖 sugar	12%	77
蛋 egg	8%	51
蘋果茸 apple sauce	28%	179
鮮酵母 fresh yeast	1%	6
肉桂粉 ground cinnamon	少許	2
無鹽牛油 unsalted butter	8%	51

自製蘋果茸	Apple sauce
蘋果 1公斤	1 kg apples
水 150克	150 g water
砂糖 30克	30 g sugar
鹽少許	salt
檸檬 1/3 個	1/3 lemon

炒蘋果粒	Apple filling
去皮青蘋果 320克	320 g peeled Granny Smith or Golden Delicious apples
肉桂粉 1/2茶匙	1/2 tsp ground cinnamon
砂糖 100克	100 g sugar
檸檬(榨汁) 1/2個	1/2 lemon (juiced)
麵粉 2湯匙	2 tbsps cake flour
牛油 30克	30 g butter
提子(用冧酒浸透)適量	Rum soaked raisins

做法 method

自製蘋果茸

1. 蘋果切成四瓣，連皮，去核。

2. 將水、檸檬和蘋果放入小鍋內，蓋上蓋子，煮至蘋果變軟，棄去檸檬，下糖、鹽調味。用攪拌機攪爛，隔去蘋果皮。

炒蘋果粒

1. 將蘋果切成大粒，擠入檸檬汁，拌勻。

2. 在不鏽鋼煎鍋內以牛油起鑊，放入砂糖，輕輕炒至砂糖微焦，下蘋果，炒至蘋果水分收乾至一半，下肉桂粉、麵粉，煮至濃稠，加入提子，關火。

Apple sauce

1. Quarter the apples with skin on. Core them.

2. Cook apples and lemon in water until tender. Remove the lemon. Season with salt and sugar. Puree until fine. Strain to remove the skin.

Apple filling

1. Cut apples into chunks. Toss in lemon juice.

2. In a stainless steel pan, melt the butter then add sugar. When the sugar starts to caramelize, put in the apples. Stir-fry until the juice reduces to half. Add ground cinnamon and flour. Cook until the juice thickens. Add raisins and remove from heat. Leave it to cool.

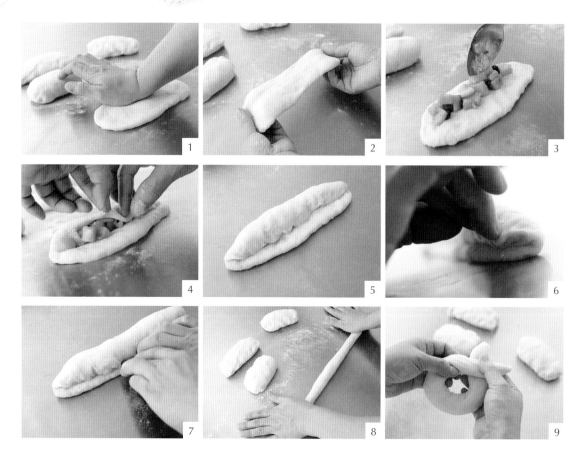

做法
method

預備

- 將中種切成小塊

混合：　　將麵糰所有材料(牛油除外)混合，搓揉，逐少加入中種，搓至柔滑，加入牛油
　　　　　再搓至可伸延薄膜。

發酵：　　麵糰放入大碗內，蓋上保鮮紙進行第一次發酵，約25-30分鐘。

分割：　　將麵糰分成十四等份，排氣，按扁，輕捲成條狀，靜置20分鐘讓麵糰鬆弛。

造型：　　將麵糰排氣，拍成長條，舀入炒蘋果粒，將麵皮覆向中央(要蓋過餡料)，揑
　　　　　實；再覆，揑實，收口向下。用掌心將麵糰搓成棒狀，捲成螺旋狀，放入模具
　　　　　內，輕輕蓋上矽膠篩或不沾布，在上面再蓋烤盤，用來控制形狀(看圖1-15)。

最後發酵：待麵糰發酵至兩倍大(看圖16)，拿開烤盤和矽膠篩，塗上蛋液，放入已預熱
　　　　　170-180℃焗爐內焗約20-25分鐘至金黃。

Preparation

- Cut the pre-ferment dough into small pieces.

1. Knead all ingredients of dough together (except the butter). Add pre-ferment dough piece by piece. Knead after each addition until soft and smooth. Add butter. Knead until stretchable consistency.

2. Put the dough into a big bowl. Cover with a cling wrap. Let it prove for about 25-30 minutes.

3. Divide the dough into fourteen equal portions and hand square each piece. Set aside to rest for about 20 minutes.

4. Flatten each piece of dough with your hands to drive the air out. Pat the dough into a rectangle. Spoon some apple filling on the rectangular dough. Fold 1/3 of the length towards the centre to cover all filling. Press the seam firmly. Then fold again. Press the seam firmly again. Turn the seam to face downward. Roll the dough gently into an elongated shape with your palms. Then coil it into a spiral shape. Put it into a mould. Cover with silicon mat or silipat and top with a baking tray to hold the shape (see pictures 1-15).

5. Let it prove again until they double in size (see picture 16). Remove the baking tray and silicon mat. Brush egg wash on top. Bake in pre-heated oven at 170-180°C for about 20-25 minutes.

貼士
tips

- 蘋果宜用青蘋果或金蘋果才爽脆；因為包餡後還需要入爐烘焙，餡料不可煮得太軟。

- Use Granny Smith or Golden Delicious for their lovely crunch and tartness. Do not overcook the filling as the buns have to be baked again in the oven.

日本黑糖棗子麵包

Black Sugar and Date Bread

材料 ingredients

中種 Pre-ferment Dough	百分比%	克 gram
筋粉 bread flour	70%	405
水 water	40%	232
鮮酵母 fresh yeast	2%	12
海鹽 sea salt	1%	6
脫脂奶粉 skim milk powder	2%	12

酵母溶於水中,再加入筋粉、海鹽、脫脂奶粉搓至柔滑,用保鮮紙包裹麵糰,放入雪櫃發酵17小時。

Dissolve the yeast in water. Add bread flour, sea salt, skim milk powder and knead until soft. Cover the dough in cling wrap. Refrigerate to let it prove for 17 hours.

麵糰 Dough	百分比%	克 gram
筋粉 bread flour	30%	174
脫脂奶粉 skim milk powder	4%	23
海鹽 sea salt	1%	6
日本黑糖 kuro sato (Japanese black sugar)	20%	116
水 water	16%	93
蛋 egg	6%	35
鮮酵母 fresh yeast	1%	6
無鹽牛油 unsalted butter	10%	58
棗子 date	25%	145

預備
preparation

- 將麵糰內的日本黑糖和水煮成糖漿，待涼備用；將棗子切粒

- 將中種切成小塊

- For the dough, cook kuro sato and water until syrupy. Set aside for later use. Dice the dates.

- Cut the pre-ferment dough into small pieces.

做法
method

混合：　　將麵糰所有材料(牛油、棗子除外)混合，搓揉，逐少加入中種，搓至柔滑，加入牛油、棗子再搓至可伸延薄膜。

發酵：　　麵糰放入大碗內，蓋上保鮮紙進行第一次發酵，約25-30分鐘。

分割：　　麵糰分割成兩等份，排氣，滾圓，蓋上保鮮紙，靜置20分鐘讓麵糰鬆弛。

造型：　　將麵糰排氣，滾圓，放入兩個聖誕麵包模內，輕按麵糰，讓麵糰更加貼近模底，用保鮮紙蓋好(看圖1-5)。

最後發酵：麵糰發酵約45分鐘至八成滿(看圖6)，蓋上不黏布或矽膠蓆，再蓋上烤盤，放入已預熱170-180℃焗爐，焗約30-35分鐘至金黃。

1. Knead all ingredients of dough together (except the butter and dates). Add pre-ferment dough piece by piece. Knead after each addition until soft and smooth. Add butter and dates. Knead until stretchable consistency.

2. Put the dough into a big bowl. Wrap with cling film and let it prove for about 25-30 minutes.

3. Divide the dough into two equal portions. Flatten each portion with your hands to drive the air out. Round it. Cover with cling wrap and set aside to rest for about 20 minutes.

4. Flatten each piece of dough with your hands to drive the air out. Round each of them and press each into a Panadora mould. Press the dough gently to fit it to the bottom of the mould. Cover with cling wrap (see pictures 1-5).

5. Let it prove for about 45 minutes or until the dough has risen to 80% of the depth of the mould (see picture 6). Cover with silicon mat and then a baking tray. Bake in a pre-heated oven at 170-180°C for about 30-35 minutes until golden brown.

tips

- 深色的糖類，烘焙時會很快上色，但注意要調低爐溫，否則會容易燒焦。
- 棗子以以色列的最優質，有去核和有核兩種，亦有有機品種，在各大型超市有售。
- Whenever brown sugar (including raw cane sugar, kuro sato, muscovado etc) is used in a dough, it tends to get browned very easily in oven. Make sure you bake it at a lower temperature. Otherwise, the bread may get burnt.
- The best dates come from Israel and they are available pitted or with pits in. You may also get organic dates in major supermarkets.

紅蘿蔔方包
Carrot Loaf

中種 Pre-ferment Dough	百分比%	克 gram
筋粉 bread flour	70%	451
水 water	40%	258
鮮酵母 fresh yeast	2%	13
海鹽 sea salt	1%	6
脫脂奶粉 skim milk powder	2%	13

酵母溶於水中,再加入筋粉、海鹽、脫脂奶粉搓至柔滑,用保鮮紙包裹麵糰,放入雪櫃發酵17小時。

Dissolve the yeast in water. Add bread flour, sea salt, skim milk powder and knead until soft. Cover the dough in cling wrap. Refrigerate to let it prove for 17 hours.

麵糰 Dough	百分比%	克 gram
筋粉 bread flour	30%	193
海鹽 sea salt	1%	6
砂糖 sugar	10%	64
紅蘿蔔汁 carrot juice	29%	187
鮮酵母 fresh yeast	1%	6
無鹽牛油 unsalted butter	7%	45
紅蘿蔔絲 julienne carrot	12%	77

做法
method

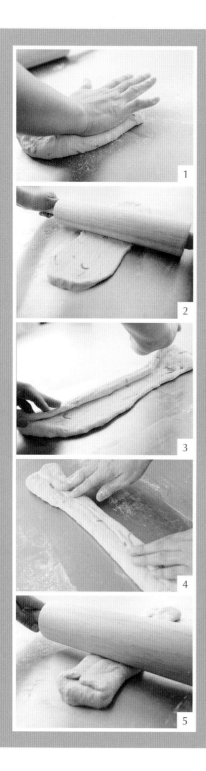

預備

• 將中種切成小塊

混合： 將麵糰所有材料(牛油、紅蘿蔔絲除外)混合，搓揉，逐少加入中種搓至柔滑，加入牛油、紅蘿蔔絲再搓至可伸延薄膜。

發酵： 麵糰放入大碗內，蓋上保鮮紙進行第一次發酵，約25-30分鐘。

分割： 將麵糰分成四等份，排氣。每份麵糰捲緊成條狀，靜置20分鐘讓麵糰鬆弛。

造型： 將麵糰排氣，再擀薄成長方形，兩邊向內摺起，再擀薄，寬度與吐司模相若。捲起後放入吐司模，不要把蓋全蓋上，用保鮮紙蓋好(看圖1-9)。

最後發酵： 麵糰發酵約45分鐘至八成滿(看圖10)，把蓋蓋上，放入已預熱170-180℃焗爐內，焗約30-35分鐘至金黃。

Preparation

- Cut the pre-ferment dough into small pieces.

1. Knead all ingredients of dough together (except the butter and julienne carrot). Add pre-ferment dough piece by piece. Knead after each addition until soft and smooth. Add butter and julienne carrot. Knead until stretchable consistency.

2. Put the dough into a big bowl. Cover with cling wrap. Let it prove for about 25-30 minutes.

3. Divide the dough into four small equal portions. Hand square each of them. Set aside to rest for about 20 minutes.

4. Flatten each piece of dough to drive the air out. Roll each piece into a rectangle. Fold both edges towards the centre and roll it out again. Roll the dough up and place dough into the mould with the seam side down without closing the lid. Cover with cling wrap (see pictures 1-9).

5. Let it prove for about 45 minutes or until the dough has risen to 80% of the depth of the mould (see picture 10). Close the lid. Bake in a pre-heated oven at 170-180°C for about 30-35 minutes.

tips

- 動一動腦筋，就可不必為某種造型的麵包買專門的模具。像圖中這小方形吐司，以小方形慕士模、蓋上大烤盤製成。

- If you use your resources flexibly, you don't have to get moulds in special shapes from baking supplies. For instance, the mini loaves shown in the picture are made with a square mousse mould covered by a baking tray.

6

7

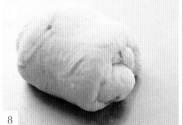

8

9

10

粟米麵包
Sweet Corn Bread

材料 *ingredients*

中種 Pre-ferment Dough	百分比%	克 gram
筋粉 bread flour	70%	403
水 water	40%	231
鮮酵母 fresh yeast	2%	12
海鹽 sea salt	1%	6
脫脂奶粉 skim milk powder	2%	12

酵母溶於水中，再加入筋粉、海鹽、脫脂奶粉搓至柔滑，用保鮮紙包裹麵糰，放入雪櫃發酵17小時。

Dissolve the yeast in water. Add bread flour, sea salt, skim milk powder and knead until soft. Cover the dough in cling wrap. Refrigerate to let it prove for 17 hours.

麵糰 Dough	百分比%	克 gram
筋粉 bread flour	30%	173
脫脂奶粉 skim milk powder	3%	17
海鹽 sea salt	1%	6
砂糖 sugar	12%	69
新鮮粟米茸 pureed fresh corn kernels	30%	173
鮮酵母 fresh yeast	1%	6
無鹽牛油 unsalted butter	7%	40
罐頭粟米粒 canned corn kernels	30%	173

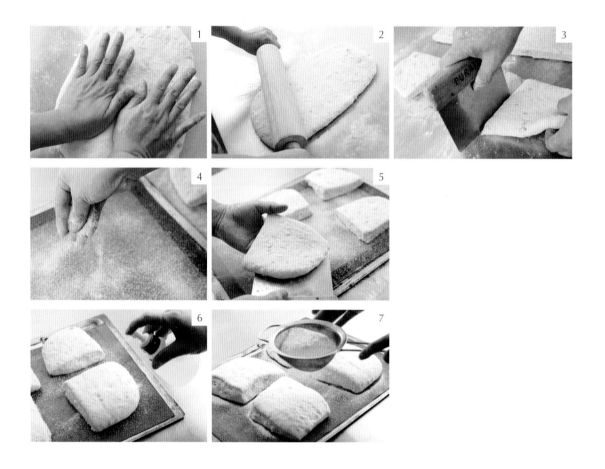

預備
preparation

- 將新鮮粟米粒放攪拌器內攪成粟米茸；瀝乾罐頭粟米粒水分

- 將中種切成小塊

- Puree fresh sweet corn kernels in a blender. Drain the canned corn kernels.

- Cut the pre-ferment dough into small pieces.

混合： 將麵糰所有材料(牛油、罐頭粟米粒除外)混合，搓揉，逐少加入中種，搓至柔滑，加牛油、罐頭粟米粒再搓至可伸延薄膜。

發酵： 麵糰放入大碗內，蓋上保鮮紙進行第一次發酵，約25-30分鐘。

鬆弛： 將麵糰按扁排氣，再擀薄成長方形，輕捲成條狀，靜置15分鐘讓麵糰鬆弛。

造型： 將麵糰按扁排氣，擀薄成長方形，隨意切成六至八塊 (看圖1-3)。

最後發酵：在焗盤上灑粟米粉，放上麵糰，發酵約45分鐘至兩倍大，噴水，篩上筋粉或其他麵粉裝飾，放入已預熱170-180℃焗爐，焗約25-30分鐘至金黃(看圖4-7)。

1. Knead all ingredients of dough together (except the butter and whole corn kernels). Add pre-ferment dough piece by piece. Knead after each addition until soft and smooth. Add butter and whole corn kernels. Knead until stretchable consistency.

2. Put the dough into a big bowl. Cover with a cling wrap. Let it prove for about 25-30 minutes.

3. Hand square the dough. Set aside to rest for about 15 minutes.

4. Flatten the dough to drive the air out. Roll it out a rectangular shape with a rolling pin. Cut into six to eight uneven pieces (see pictures 1-3).

5. Sprinkle some cornmeal on a baking tray. Put the dough in it. Let it prove for about 45 minutes or until the dough has doubled in size. Spray some water on the dough. Sift some flour over it. Bake in a pre-heated oven at 170-180°C for about 25-30 minutes (see pictures 4-7).

貼士
tips

- 灑在焗盤上的黃色小粒狀粟米粉，在大型超市或烘焙用品專門店有售。

- 把粟米粉灑在焗盤上才放上麵包作最後發酵，可使麵包底部更香脆；但也可以筋粉代替。

- Cornmeal is available from major supermarkets or baking supply stores.

- Dusting the baking tray with cornmeal before putting in the dough for the second rise helps make the bottom crust more crispy. You may also use bread flour instead if you can't get it.

香芋麵包
Taro Bun

材料
ingredients

中種 Pre-ferment Dough	百分比%	克 gram
筋粉 bread flour	70%	438
水 water	40%	250
鮮酵母 fresh yeast	2%	13
海鹽 sea salt	1%	6
脱脂奶粉 skim milk powder	2%	13

酵母溶於水中，再加入筋粉、海鹽、脱脂奶粉搓至柔滑，用保鮮紙包裹麵糰，放入雪櫃發酵17小時。

Dissolve the yeast in water. Add bread flour, sea salt, skim milk powder and knead until soft. Cover the dough in cling wrap. Refrigerate to let it prove for 17 hours.

麵糰 Dough	百分比%	克 gram
筋粉 bread flour	30%	188
脱脂奶粉 skim milk powder	4%	25
海鹽 sea salt	1%	6
砂糖 sugar	12%	75
水 water	14%	88
蛋 egg	8%	50
鮮酵母 fresh yeast	1%	6
無鹽牛油 unsalted butter	6%	38
蒸熟芋頭絲 steamed shredded taro	20%	125

餡料：	Filling：
蒸熟芋頭絲220克	220 g steamed shredded taro
糖霜55克	55 g icing sugar
*拌勻	*mixed well

預備
preparation

- 芋頭刨絲蒸熟，預留一些用來做餡
- 將中種切成小塊
- Shred the taro and steam it. Save part of it for filling.
- Cut the pre-ferment dough into small pieces.

做法
method

混合：	將麵糰所有材料(牛油、蒸熟芋頭絲20%除外)混合，搓揉，逐少加入中種，搓至柔滑，加入牛油和蒸熟芋頭絲20%，再搓至可伸延薄膜。
發酵：	麵糰放入大碗內，蓋上保鮮紙進行第一次發酵，約25-30分鐘。
分割：	將麵糰分成十六等份，滾圓，靜置20分鐘讓麵糰鬆弛。
造型：	將麵糰排氣，包入餡料，八個一組放入長方形淺盤內，用保鮮紙蓋好(看圖1-7)。
最後發酵：	麵糰發酵約45分鐘至八成滿(看圖8)，塗上蛋液，放入已預熱170-180℃焗爐，焗約30-35分鐘至金黃。

1. Knead all ingredients of dough together (except the butter and 20% of the steamed taro). Add pre-ferment dough piece by piece. Knead after each addition until soft and smooth. Add butter and 20% of the steamed taro. Knead until stretchable consistency.

2. Put the dough into a big bowl. Cover with a cling wrap. Let it prove for about 25-30 minutes.

3. Divide the dough into sixteen equal portions. Round them. Set aside to rest for about 20 minutes.

4. Flatten each piece of dough with your hands to drive the air out. Stuff them with taro filling. Arrange eight pieces of stuffed dough into one mould. Cover with cling wrap (see pictures 1-7).

5. Let it prove for about 45 minutes or until the dough has risen to 80% of the depth of the mould (see picture 8). Brush egg wash on top. Bake in a pre-heated oven at 170-180°C for about 30-35 minutes.

貼士
tips

- 不要使用小芋頭，要買荔浦芋或泰國芋才會粉。在中秋前後買到的中國芋頭都不俗，雖然泰國芋頭價格比中國芋頭貴，但質量穩定。

- Do not use baby taros for this recipe. Get Lei Po taros or Thai taros because they are more starchy. Those Chinese taros available around mid-autumn festival are also good. Although Thai taros are more expensive, they are more consistent in starch content.

油浸半乾番茄麵包
Semi Sun-dried Tomato Bread

中種 Pre-ferment Dough	百分比%	克 gram
筋粉 bread flour	70%	407
水 water	40%	233
鮮酵母 fresh yeast	2%	12
海鹽 sea salt	1%	6
脫脂奶粉 skim milk powder	2%	12

酵母溶於水中，再加入筋粉、海鹽、脫脂奶粉搓至柔滑，用保鮮紙包裹麵糰，放入雪櫃發酵17小時。

Dissolve the yeast in water. Add bread flour, sea salt, skim milk powder and knead until soft. Cover the dough in cling wrap. Refrigerate to let it prove for 17 hours.

麵糰 Dough	百分比%	克 gram
筋粉 bread flour	30%	175
脫脂奶粉 skim milk powder	2%	12
海鹽 sea salt	0.8%	4
砂糖 sugar	12%	70
蜂蜜 honey	4%	23
蛋 egg	8%	47
罐裝去皮番茄 Italian peeled plum tomatoes (canned)	16%	93
番茄膏 tomato paste	20%	116
鮮酵母 fresh yeast	1%	6
橄欖油 olive oil	8%	47
油浸半乾番茄 semi sun-dried tomato	10%	58

做法
method

預備

● 將中種切成小塊

● 油浸半乾番茄切粒，隔去罐裝去皮番茄汁液，待用

混合： 將麵糰所有材料(油浸半乾番茄除外)混合，搓揉，逐少加入中種，搓至柔
滑，加入油浸半乾番茄搓至可伸延薄膜。

發酵： 麵糰放入大碗內，蓋上保鮮紙進行第一次發酵，約25-30分鐘。

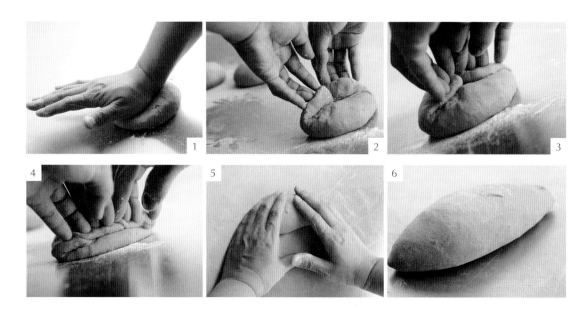

分割： 將麵糰分成八等份，排氣，滾圓，靜置20分鐘讓麵糰鬆弛。

造型： 將麵糰排氣，捲成欖核形，四個一組放入吐司模內發酵，輕輕蓋上保鮮紙
(看圖1-10)。

最後發酵：麵糰發酵約30分鐘至八成滿，塗上蛋液，中間擠一條牛油，放入已預熱170-
180℃焗爐，焗約30-35分鐘至金黃(看圖11-12)。

貼士

● 油浸半乾番茄以澳洲出產的為佳，味道較鮮甜，在大型超級市場有售。

Preparation

- Cut the pre-ferment dough into small pieces.
- Dice the semi sun-dried tomato. Drain the Italian peeled plum tomatoes.

1. Knead all ingredients of dough together (except the semi sun-dried tomatoes). Add pre-ferment dough piece by piece. Knead after each addition until soft and smooth. Add semi sun-dried tomatoes. Knead until stretchable consistency.
2. Put the dough into a big bowl. Cover with a cling wrap and let it prove for about 25-30 minutes.

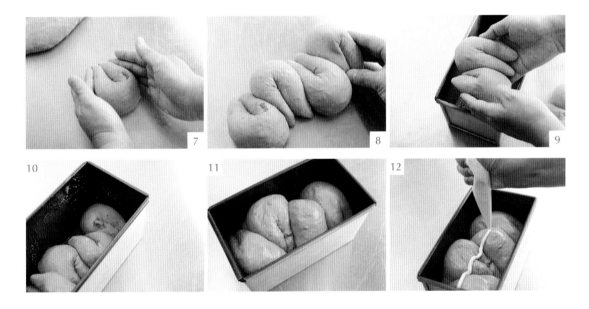

3. Divide the dough into eight equal portions and round it. Set aside to rest for about 20 minutes.
4. Flatten each piece of dough with your hands to drive the air out. Roll each into a long oval shape. Fold up the dough from one end to another. Place four pieces of dough into a loaf tin. Cover with cling wrap (see pictures 1-10).
5. Let it prove for about 30 minutes or until the dough has risen to 80% of the depth of the mould. Brush egg wash on top. Pipe a line of softened butter at the centre of dough. Bake in a pre-heated oven at 170-180°C for about 30-35 minutes (see pictures 11-12).

Tips

- Semi sun-dried tomatoes are usually steeped in olive oil. Those originated from Australia are known to be of the best quality with a sweeter taste. They are available from major supermarkets.

杏仁香蕉麵包
Almond Banana Bread

中種 Pre-ferment Dough	百分比%	克 gram
筋粉 bread flour	70.0%	445
水 water	40.0%	254
鮮酵母 fresh yeast	2.0%	13
海鹽 sea salt	1.0%	6
脫脂奶粉 skim milk powder	2.0%	13

酵母溶於水中，再加入筋粉、海鹽、脫脂奶粉搓至柔滑，用保鮮紙包裹麵糰，放入雪櫃發酵17小時。

Dissolve the yeast in water. Add bread flour, sea salt, skim milk powder and knead until soft. Cover the dough in cling wrap. Refrigerate to let it prove for 17 hours.

麵糰 Dough	百分比%	克 gram
筋粉 bread flour	30%	191
脫脂奶粉 skim milk powder	3%	19
鹽 salt	0.5%	3
鮮酵母 fresh yeast	1%	6
砂糖 sugar	12%	76
蛋 egg	6%	38
熟香蕉 riped banana	30%	191
無鹽牛油 unsalted butter	10%	64
香草莢 vanilla pod		一枝 1pc

杏仁醬	Frangipane filling	裝飾	Decoration
杏仁粉 159克	159 g ground almond	香蕉 4條	4 bananas
糖 95克	95 g sugar	糖 適量	sugar
牛油 159克	159 g butter	檸檬汁 少許	lemon juice
蛋 159克	159 g eggs	肉桂粉 適量	ground cinnamon
麵粉 16克	16 g flour		
冧酒 13克	13 g Rum		

做法
method

預備

- 將中種切成小塊
- 糖和肉桂粉調勻
- 香蕉切片，灑上檸檬汁以防變黑
- 剝開香草莢，用刀刮取種籽，待用

混合： 將麵糰所有材料（牛油除外）混合，搓揉，逐少加入中種，搓至柔滑，加入牛油，再搓至可伸延薄膜。

發酵： 麵糰放入大碗內，蓋上保鮮紙進行第一次發酵，約25-30分鐘。

分割： 將麵糰分成四等份，排氣，按扁，輕捲成條狀，靜置20分鐘讓麵糰鬆弛(看圖1)。

造型： 將麵糰擀薄成長方形，放上長方形餅模，切去多餘部分，擠上杏仁醬；香蕉片沾上肉桂糖，排在麵糰上，用保鮮紙蓋好待發酵(看圖2-8)。

最後發酵：待麵糰發酵至兩倍大，放入已預熱170-180℃的焗爐內，焗約20-25分鐘至金黃。

杏仁醬

牛油和糖用打蛋機打發至奶白色，蛋分幾次加入，拌入麵粉、杏仁粉和冧酒，攪勻待用。

Preparation

- Cut the pre-ferment dough into small pieces.
- Mix ground cinnamon with sugar.
- Slice the bananas and toss them in lemon juice to avoid rusting.
- Cut open the vanilla pod and scrape out the seeds.

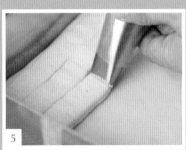

1. Knead all ingredients of dough together (except the butter). Add pre-ferment dough piece by piece. Knead after each addition until soft and smooth. Add butter. Knead until stretchable consistency.

2. Put the dough into a big bowl. Cover with a cling wrap. Let it prove for about 25-30 minutes.

3. Divide the dough into four equal pieces. Flatten with your hands to drive the air out. Press them flat again and hand square them. Set aside to rest for about 20 minutes (see picture 1).

4. Roll each dough out into a rectangle and put the mould on. Trim off the edge and pipe almond filling on top. Coat the banana slices in cinnamon sugar. Arrange over the dough. Cover with cling wrap (see pictures 2-8).

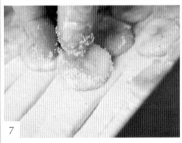

5. Let it prove until the dough doubles in size. Bake in a pre-heated oven at 170-180°C for about 20-25 minutes.

Frangipane filling

Cream the butter with sugar until pale. Gradually beat in the eggs. Fold in ground almond, cake flour and Rum.

紫心番薯麵包
Purple Sweet Potato Bread

中種 Pre-ferment Dough	百分比%	克 gram
筋粉 bread flour	70%	377
水 water	40%	216
鮮酵母 fresh yeast	2%	11
海鹽 sea salt	1%	5
脫脂奶粉 skim milk powder	2%	11

酵母溶於水中，再加入筋粉、海鹽、脫脂奶粉搓至柔滑，用保鮮紙包裹麵糰，放入雪櫃發酵17小時。

Dissolve the yeast in water. Add bread flour, sea salt, skim milk powder and knead until soft. Cover the dough in cling wrap. Refrigerate to let it prove for 17 hours.

麵糰 Dough	百分比%	克 gram
筋粉 bread flour	30%	162
脫脂奶粉 skim milk powder	3%	16
海鹽 sea salt	1%	5
砂糖 sugar	12%	65
水 water	25%	135
鮮酵母 fresh yeast	1%	5
無鹽牛油 unsalted butter	8%	43
紫心番薯茸 mashed purple sweet potato	50%	269

餡料	Filling
紫心番薯茸 適量	mashed purple sweet potato

預備
preparation

- 將中種切成小塊

- 紫心番薯連皮蒸熟，起肉，壓成茸，預留一些用來做餡

- Cut the pre-ferment dough into small pieces.

- Steam the purple sweet potato with the skin on. Peel and mash it. Save some for filling.

混合： 將麵糰所有材料(牛油除外)混合，搓揉，逐少加入中種，搓至柔滑，加入牛油再搓至可伸延薄膜。

發酵： 麵糰放入大碗內，蓋上保鮮紙進行第一次發酵，約25-30分鐘。

分割： 將麵糰分成十六等份，排氣，滾圓，靜置20分鐘讓麵糰鬆弛。

造型： 將麵糰再次排氣，滾圓，壓扁，包入紫心番薯茸，捏緊收口，再滾圓，八個一組放入雪芳蛋糕模，用保鮮紙蓋好(看圖1-8)。

最後發酵： 麵糰發酵約45分鐘，發起至八成滿(看圖9)，塗上蛋水，放入已預熱170-180℃焗爐內，焗約30-35分鐘至金黃。

1. Knead all ingredients of dough together (except the butter). Add pre-ferment dough piece by piece. Knead after each addition until soft and smooth. Add butter. Knead until stretchable consistency.

2. Put the dough into a big bowl. Cover with a cling wrap. Let it prove for about 25-30 minutes.

3. Divide the dough into sixteen small equal portions and round them. Set aside to rest for about 20 minutes.

4. Flatten each piece of dough with your hands to drive air out. Round them again. Press flat. Stuff them with mashed sweet potato. Round again. Place eight pieces of dough into one mould. Cover with cling wrap (see pictures 1-8).

5. Let it prove for about 45 minutes or until the dough has risen to 80% of the depth of the mould (see picture 9). Brush egg wash on top. Bake in a pre-heated oven at 170-180°C for about 30-35 minutes.

貼士
tips

- 使用深紫色的番薯方能做到圖中色澤；但也可以用黃心或紅心番薯代替。

- Purple sweet potato is used because of the sheen it creates on the bread (as shown in the picture). Of course you may also use yellow or orange sweet potato instead.

香橙朱古力麵包
Orange Chocolate Chip Bread

中種 Pre-ferment Dough	百分比%	克 gram
筋粉 bread flour	70%	462
水 water	40%	264
鮮酵母 fresh yeast	2%	13
海鹽 sea salt	1%	7
脫脂奶粉 skim milk powder	2%	13

酵母溶於水中，再加入筋粉、海鹽、脫脂奶粉搓至柔滑，用保鮮紙包裹麵糰，放入雪櫃發酵17小時。

Dissolve the yeast in water. Add bread flour, sea salt, skim milk powder and knead until soft. Cover the dough in cling wrap. Refrigerate to let it prove for 17 hours.

麵糰 Dough	百分比%	克 gram
筋粉 bread flour	30%	198
脫脂奶粉 skim milk powder	3%	20
海鹽 sea salt	1%	7
砂糖 sugar	10%	66
水 water	2%	13
蛋 egg	6%	40
香橙果醬 orange marmalade	24%	158
鮮酵母 fresh yeast	1%	7
無鹽牛油 unsalted butter	8%	53

餡料	filling
入爐朱古力粒100克	100 g baking chocolate chips
橙皮(磨茸)(或適量蜜餞橙皮)2個	2 pc grated orange zest or glaced orange peel

預備
preparation

- 將中種切成小塊
- Cut the pre-ferment dough into small pieces.

做法
method

混合： 將麵糰所有材料(牛油、入爐朱古力粒和蜜餞橙皮除外)混合，搓揉，逐少加入中種，搓至柔滑，加入牛油、入爐朱古力粒和蜜餞橙皮搓至可伸延薄膜。

發酵： 麵糰放入大碗內，蓋上保鮮紙進行第一次發酵，約25-30分鐘。

分割： 將麵糰分成三十六等份，排氣，滾圓，靜置20分鐘讓麵糰鬆弛(看圖1-4)。

造型： 拍出麵糰空氣，滾成圓形，九個一組放入吐司模內，輕輕蓋上保鮮紙(看圖5-7)。

最後發酵： 麵糰發酵約40-45分鐘至八成滿，塗上蛋液，放入已預熱170-180℃爐焗內焗約30-35分鐘至金黃。

1. Knead all ingredients together (except the butter, chocolate chips and the orange zest). Add pre-ferment dough piece by piece. Knead after each addition until soft and smooth. Add butter, chocolate chips and the orange zest. Knead until stretchable consistency.

2. Put the dough into a big bowl. Cover with a cling wrap. Let it prove for 25-30 minutes.

3. Divide the dough into thirty six small equal portions. Flatten to drive out the air. Roll them into balls. Set aside to rest for about 20 minutes (see pictures 1-4).

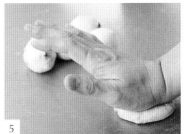

4. Flatten and roll them round again. Place nine dough balls into a loaf tin. Cover with cling wrap (see pictures 5-7).

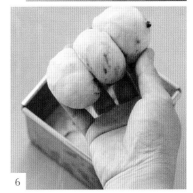

5. Let it prove again for about 40-45 minutes or until the dough has risen to 80% of the depth of the loaf tin. Brush egg wash on top. Bake in a pre-heated oven at 170-180°C for about 30-35 minutes.

貼士
tips

- 入爐朱古力粒經高溫烘焙後也不易溶化。
- Baking chocolate can stand high heat. It doesn't melt in the baking process.

鳳梨甜包
Buns with Pineapple Filling

中種 Pre-ferment Dough	百分比%	克 gram
筋粉 bread flour	70%	476
水 water	40%	272
鮮酵母 fresh yeast	2%	14
海鹽 sea salt	1%	7
脫脂奶粉 skim milk powder	2%	14

酵母溶於水中，再加入筋粉、海鹽、脫脂奶粉搓至柔滑，用保鮮紙包裹麵糰，放入雪櫃發酵17小時。

Dissolve the yeast in water. Add bread flour, sea salt, skim milk powder and knead until soft. Cover the dough in cling wrap. Refrigerate to let it prove for 17 hours.

麵糰 Dough	百分比%	克 gram
筋粉 bread flour	30%	204
脫脂奶粉 skim milk powder	3%	20
海鹽 sea salt	1%	7
砂糖 sugar	15%	102
水 water	9%	61
蛋 egg	10%	68
鮮酵母 fresh yeast	1%	7
無鹽牛油 unsalted butter	10%	68

餡料	Filling
鳳梨果醬適量	homemade pineapple jam
*請看第142頁的製法	* Please refer to P.142 for the method.

做法
method

預備

● 將中種切成小塊

混合： 將麵糰所有材料(牛油除外)混合，搓
揉，逐少加入中種，搓至柔滑，加入牛
油再搓至可伸延薄膜。

發酵： 麵糰放入大碗內，蓋上保鮮紙進行第一
次發酵，約25-30分鐘。

分割： 麵糰分割成十五等份，排氣，滾圓，靜
置20分鐘讓麵糰鬆弛。

造型： 將麵糰排氣再滾圓，壓扁包入鳳梨果
醬，捏實收口，三個一組放入模中，用
保鮮紙蓋好(看圖1-8)。

最後發酵： 麵糰發酵約45分鐘至八成滿，塗上蛋
液，放一小片牛油在中心，放入已預熱
170-180℃的焗爐內焗約20-25分鐘至
金黃(看圖9-10)。

Preparation

- Cut the pre-ferment dough into small pieces.

1. Knead all ingredients of dough together (except the butter). Add pre-ferment dough piece by piece. Knead after each addition until soft and smooth. Add butter. Knead until stretchable consistency.

2. Put the dough into a big bowl and cover with a cling wrap. Let it prove for about 25-30 minutes.

3. Divide the dough into fifteen equal portions and roll them round. Set aside to rest for about 20 minutes.

4. Flatten the dough with your hands to drive the air out. Roll them round again. Stuff pineapple filling into each dough ball. Put three balls in one mould. Cover with a cling wrap (see pictures 1-8).

5. Let it prove for about 45 minutes or until the dough doubles in size. Brush egg wash on top. Put a little piece of butter over the dough. Bake in a pre-heated oven at 170-180°C for about 20-25 minutes (see pictures 9-10).

tips

- 放一小片牛油在麵糰中心，可令麵包的裂口更美觀。

- Put a little piece of butter at the centre of the dough makes the bread has beautiful cracks after baked.

百里香鵝肝麵包

Foie Gras and Thyme Loaf

材料
ingredients

中種 Pre-ferment Dough	百分比%	克 gram
筋粉 bread flour	70%	467
水 water	40%	267
鮮酵母 fresh yeast	2%	13
海鹽 sea salt	1%	7
脫脂奶粉 skim milk powder	2%	13

酵母溶於水中，再加入筋粉、海鹽、脫脂奶粉搓至柔滑，用保鮮紙包裹麵糰，放入雪櫃發酵17小時。

Dissolve the yeast in water. Add bread flour, sea salt, skim milk powder and knead until soft. Cover the dough in cling wrap. Refrigerate to let it prove for 17 hours.

麵糰 Dough	百分比%	克 gram
筋粉 bread flour	30%	200
脫脂奶粉 skim milk powder	3%	20
海鹽 sea salt	1%	7
砂糖 sugar	8%	53
蜂蜜 honey	3%	20
水 water	18%	120
蛋 egg	5%	33
鮮酵母 fresh yeast	1%	7
無鹽牛油 unsalted butter	2%	13
鵝或鴨肝醬 foie gras or duck liver pate	12%	80
鵝絲 goose rillettes	18%	120
新鮮百里香 fresh thyme		適量
黑椒 ground black pepper		適量

做法
method

預備

• 將中種切成小塊

混合：　將麵糰所有材料(牛油、鵝絲除外)混合，搓揉，逐少加入中種，搓至柔滑，加入牛油、鵝絲再搓至可伸延薄膜。

發酵：　麵糰放入大碗內，蓋上保鮮紙進行第一次發酵，約25-30分鐘。

分割：　麵糰分割成廿一等份，排氣，滾圓，蓋上保鮮紙，靜置20分鐘讓麵糰鬆弛。

造型：　將麵糰排氣，滾成圓形，七個一組放入已塗油模內，輕輕蓋上保鮮紙(看圖1-5)。

最後發酵：麵糰發酵約40-45分鐘至八成滿(看圖6)，塗上蛋液，放入已預熱170-180℃焗爐內，焗約30-35分鐘至金黃。

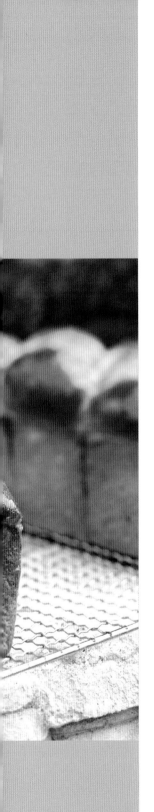

Preparation

- Cut the pre-ferment dough into small pieces.

1. Knead all ingredients together (except butter and goose rillettes). Add pre-ferment dough piece by piece. Knead after each addition until soft and smooth. Add butter and goose rillettes. Knead until stretchable consistency.

2. Put the dough into a big bowl. Cover with cling wrap and let it prove for about 25-30 minutes.

3. Divide the dough into twenty one small equal pieces. Flatten each piece of dough with your hands to drive the air out and then round it. Cover with cling wrap. Set aside to rest for about 20 minutes.

4. Flatten the dough with your hands to drive the air out. Round each piece again. Put seven pieces of round dough into a loaf tin. Cover with cling wrap (see pictures 1-5).

5. Let it prove for about 40-45 minutes or until the dough has risen to 80% of the depth of the loaf tin (see picture 6). Bake in a pre-heated oven at 170-180°C for about 30-35 minutes until golden brown.

意式風情
Mini Italian Swirls

中種 Pre-ferment Dough	百分比%	克 gram
筋粉 bread flour	70%	491
水 water	40%	281
鮮酵母 fresh yeast	2%	14
海鹽 sea salt	1%	7
脫脂奶粉 skim milk powder	2%	14

酵母溶於水中，再加入筋粉、海鹽、脫脂奶粉搓至柔滑，用保鮮紙包裹麵糰，放入雪櫃發酵17小時。

Dissolve the yeast in water. Add bread flour, sea salt, skim milk powder and knead until soft. Cover the dough in cling wrap. Refrigerate to let it prove for 17 hours.

麵糰 Dough	百分比%	克 gram
筋粉 bread flour	30%	210
海鹽 sea salt	1.2%	8
砂糖 sugar	8%	56
水 water	20%	140
鮮酵母 fresh yeast	1%	7
橄欖油 olive oil	8%	56
綠色香蒜醬 green pesto	5%	35
雜香草 mixed dried herbs		適量

餡料	Filling
已烤紅甜椒一個	1 grilled red capsicum
已烤黃甜椒一個	1 grilled yellow capsicum
已烤綠甜椒一個	1 grilled green capsicum
黑水欖適量	black olives
新鮮羅勒適量	fresh basil ⎤
銀魚柳1-2條	1-2 anchovy fillet ⎦ 調勻 mixed well
綠色香蒜醬約三湯匙	3 tbsps green pesto
巴馬臣芝士碎適量	grated parmesan cheese

羅勒香蒜醬	Green pesto
新鮮羅勒70克	70 g fresh basil
松子仁25克	25 g pine nuts
腰果25克	25 g cashew nuts
蒜頭3瓣	3 cloves garlic
海鹽適量	sea salt
巴馬臣芝士50克	50 g parmesan cheese
橄欖油120克	120 g olive oil

做法 *method*

預備

- 將中種切成小塊
- 黑水欖切片

混合： 將麵糰所有材料混合，搓揉，逐少加入中種，搓至可伸延薄膜。

發酵： 麵糰放入大碗內，蓋上保鮮紙進行第一次發酵，約25-30分鐘。

分割： 將麵糰排氣，按扁，輕捲成條狀，蓋上保鮮紙，靜置20分鐘讓麵糰鬆弛。

造型： 將麵糰排氣，再擀薄至長方形，塗上羅勒香蒜醬，放上三色甜椒條、黑水欖片、羅勒、刨上芝士碎，捲起，用牙線切成十二等份，用保鮮紙蓋好(看圖1-10)。

最後發酵： 麵糰發酵至雙倍大，放上三色甜椒、黑水欖裝飾，刨上芝士碎，放入已預熱170-180℃的焗爐內，焗約25-30分鐘至金黃(看圖11-12)。

羅勒香蒜醬

- 用食物處理器將所有材料攪拌成醬。

三色甜椒處理法

- 將三色甜椒用鉗夾着，放在爐火上燒至表皮焦黑，放入大碗內，蓋上保鮮紙，待表皮軟化，撕去皮，切條。

Preparation

- Cut the pre-ferment dough into small pieces.
- Cut black olives into slices.

1. Knead all ingredients together. Add pre-ferment dough piece by piece. Knead after each addition until stretchable consistency.
2. Put the dough into a big bowl. Cover with cling wrap and let it prove for about 25-30 minutes.
3. Flatten the dough with your hands to drive the air out. Hand square the dough. Cover with cling wrap. Set aside to rest for about 20 minutes.

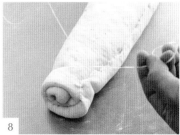

4. Flatten the dough with your hands to drive the air out. Roll the dough out into a rectangle with a rolling pin. Spread pesto on top. Arrange capsicums, black olives and basil leaves on top. Shave some parmesan cheese over it. Roll the dough up. Cut into twelve pieces with a dental floss. Cover with cling wrap (see pictures 1-10).

5. Let it prove until it doubles in size. Put more capsicums, black olives and cheese on top. Bake in a pre-heated oven at 170-180˚C for about 25-30 minutes until golden brown (see pictures 11-12).

Green pesto

- Blend all ingredients in a food processor until smooth.

Grilling capsicums

- Hold the capsicums with a pair of tongs. Burn them over naked flame until the skin is charred. Put them in a large mixing bowl and cover with cling wrap. Leave them to sweat for a while. Rub off the charred skins and julienne them.

tips

- 羅勒香蒜醬在各超市有售，緊記價格和質量是成正比的。
- Instead of making your own green pesto, you can get them in a bottle from supermarkets. You only get what you pay for -- its quality is proportional to its price. Of course, freshly made pesto always tastes best.

羅勒紅洋蔥巴馬腿包

Parma Ham Bread with Basil and Red Onion

中種 Pre-ferment Dough	百分比%	克 gram
筋粉 bread flour	70%	395
水 water	40%	225
鮮酵母 fresh yeast	2%	11
海鹽 sea salt	1%	6
脫脂奶粉 skim milk powder	2%	11

酵母溶於水中，再加入筋粉、海鹽、脫脂奶粉搓至柔滑，用保鮮紙包裹麵糰，放入雪櫃發酵17小時。

Dissolve the yeast in water. Add bread flour, sea salt, skim milk powder and knead until soft. Cover the dough in cling wrap. Refrigerate to let it prove for 17 hours.

麵糰 Dough	百分比%	克 gram
筋粉 bread flour	30%	169
脫脂奶粉 skim milk powder	3%	17
海鹽 sea salt	1.2%	7
砂糖 sugar	8%	45
水 water	20%	113
鮮酵母 fresh yeast	1%	6
橄欖油 olive oil	8%	45
巴馬腿 Parma ham	18%	101
切碎羅勒 chopped basil	5%	28
紅洋蔥 red onion	25%	141

做法
method

預備

- 將中種切成小塊
- 巴馬腿、紅洋葱切成幼粒，羅勒切碎

混合： 將麵糰所有材料(巴馬腿、紅洋葱除外)混合，搓揉，逐少加入中種，搓至柔滑，加入巴馬腿、紅洋葱再搓至可伸延薄膜。

發酵： 麵糰放入大碗內，蓋上保鮮紙進行第一次發酵，約25-30分鐘。

分割： 麵糰分割成五等份，排氣，滾圓，蓋上保鮮紙，靜置20分鐘讓麵糰鬆弛。

造型： 將麵糰排氣，滾成上圓下尖像淚滴形的麵糰，麵糰放入已塗油陶模內，輕輕蓋上保鮮紙(看圖1-6)。

最後發酵： 麵糰發酵約40-45分鐘至八成滿(看圖7)，塗上蛋液，放入已預熱170-180℃的焗爐，焗約30-35分鐘至金黃。

Preparation

- Cut the pre-ferment dough into small pieces.
- Finely chop the Parma ham, red onion and basil.

1. Knead all ingredients of dough together (except the Parma ham and red onion). Add pre-ferment dough piece by piece. Knead after each addition until soft and smooth. Add the Parma ham and red onion. Knead until stretchable consistency.

2. Put the dough into a big bowl. Cover with a cling wrap. Let it prove for about 25-30 minutes.

3. Divide the dough into five equal portions. Flatten the dough with your hands to drive the air out and round them. Cover with cling wrap. Set aside to rest for about 20 minutes.

4. Flatten the dough with your hands to drive the air out. Roll the dough into tear-drop shape (a cone with round bottom) with your palms. Put them into a terracotta pot. Cover with cling wrap (see pictures 1-6).

5. Let it prove for about 40-45 minutes or until the dough has risen to 80% of the depth of the mould (see picture 7). Brush egg wash on top. Bake in a pre-heated oven at 170-180°C for about 30-35 minutes.

抹茶红豆麵包
Matcha Red Bean Buns

材料 ingredients

中種 Pre-ferment Dough	百分比%	克 gram
筋粉 bread flour	70%	486
水 water	40%	278
鮮酵母 fresh yeast	2%	14
海鹽 sea salt	1%	7
脱脂奶粉 skim milk powder	2%	14

酵母溶於水中，再加入筋粉、海鹽、脱脂奶粉搓至柔滑，用保鮮紙包裹麵糰，放入雪櫃發酵17小時。

Dissolve the yeast in water. Add bread flour, sea salt, skim milk powder and knead until soft. Cover the dough in cling wrap. Refrigerate to let it prove for 17 hours.

麵糰 Dough	百分比%	克 gram
筋粉 bread flour	30%	208
鮮忌廉 whipping cream	10%	69
宇治抹茶粉 Matcha green tea powder	3%	21
海鹽 sea salt	1%	7
三溫糖 light brown sugar (wasanontou)	15%	104
水 water	5%	35
蛋 egg	5%	35
鮮酵母 fresh yeast	1%	7
無鹽牛油 unsalted butter	5%	35
甜紅豆粒 cooked red beans		適量

甜紅豆粒	Sweet red bean filling
大粒紅豆250克	250 g red beans
砂糖225克	225 g sugar
葡萄糖膠20克	20 g glucose
水適量	water

做法
method

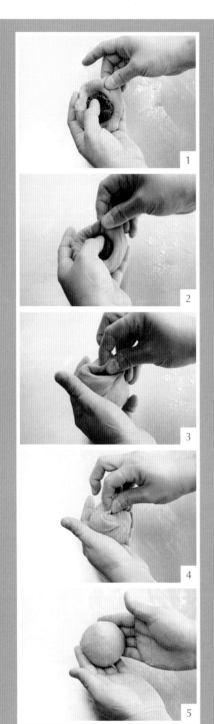

預備

- 將中種切成小塊

混合： 將麵糰所有材料(牛油、甜紅豆粒除外)混合，搓揉，逐少加入中種，搓至柔滑，加入牛油搓至可伸延薄膜。

發酵： 麵糰放入大碗內，蓋上保鮮紙，進行第一次發酵，約25-30分鐘。

分割： 將麵糰分成每個80克等份，排氣，滾圓，蓋上保鮮紙，靜置20分鐘讓麵糰鬆弛。

造型： 將麵糰排氣，滾成圓形，拍扁，包入甜紅豆粒(約22克)，收口向下，放在焗盤上，輕輕蓋上保鮮紙，待發酵(看圖1-6)。

最後發酵：麵糰發酵至兩倍大，塗上蛋液，放入已預熱170-180℃焗爐，焗約25-30分鐘至金黃；燒紅烙模，在包上烙花紋(看圖7-10)。

甜紅豆粒製法

1. 紅豆洗淨，用清水浸一晚。瀝去水分，下水以大火煮一小時，不時攪拌，但不可大力，以免壓爛紅豆。瀝去水分，以大鍋蒸45分鐘。

2. 紅豆和糖一起放鍋內煮至糖溶，下葡萄糖膠，以慢火煮至水分收乾和紅豆變軟。

貼士

- 葡萄糖膠可令紅豆有光澤，如找不到可以省卻。可買罐裝紅豆茸代替自煮甜紅豆粒。

Preparation

- Cut the pre-ferment dough into small pieces.

1. Knead all ingredients of dough together (except the butter and sweet red beans). Add pre-ferment dough piece by piece. Knead after each addition until soft and smooth. Add butter. Knead until stretchable consistency.

2. Put the dough into a big bowl. Cover with a cling wrap. Let it prove for about 25-30 minutes.

3. Divide the dough into equal portions weighing 80 g each. Flatten each piece of dough with your hands to drive the air out and then round it. Cover with cling wrap. Set aside to rest for about 20 minutes.

4. Flatten the dough with your hands to drive the air out. Round them and press flat. Stuff each with sweet red bean filling (about 22 g each). Seal and place them on a baking tray with the seam side down. Cover with cling wrap (see pictures 1-6).

5. Let it prove until they double in size. Brush egg wash on top. Bake in a pre-heated oven at 170-180°C for about 25-30 minutes until golden brown. Heat the metal pattern stamp and cauterize the pattern on the buns (see pictures 7-10).

Sweet red bean filling

1. Soak the red beans in water overnight. Drain. Cook red beans in water over high heat for 1 hour. Stir occasionally but don't mash the beans. Drain the water. Steam the red beans for 45 minutes.

2. Cook red beans and sugar until the sugar melts. Add glucose and cook over low heat until the filling turns shiny, the beans are soft and the liquid almost dries out.

Tips

- Glucose makes the filling shinier. If you can't get it, you may skip it. You may also use canned Japanese Azuki beans instead of making your own filling.

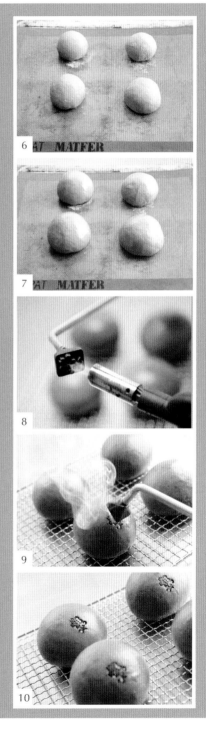

椰汁斑蘭卷
Pandan Coconut Twist Bread

Tender relationship between earth and

82

中種 Pre-ferment Dough	百分比%	克 gram
筋粉 bread flour	70%	459
水 water	40%	262
鮮酵母 fresh yeast	2%	13
海鹽 sea salt	1%	7
脫脂奶粉 skim milk powder	2%	13

酵母溶於水中，再加入筋粉、海鹽、脫脂奶粉搓至柔滑，用保鮮紙包裹麵糰，放入雪櫃發酵17小時。

Dissolve the yeast in water. Add bread flour, sea salt, skim milk powder and knead until soft. Cover the dough in cling wrap. Refrigerate to let it prove for 17 hours.

麵糰 Dough	百分比%	克 gram
筋粉 bread flour	30%	197
脫脂奶粉 skim milk powder	3%	20
海鹽 sea salt	1%	7
砂糖 sugar	15%	98
蛋 egg	4%	26
斑蘭葉汁 pandan juice	16%	105
椰汁 coconut milk	8.5%	56
鮮酵母 fresh yeast	1%	7
無鹽牛油 unsalted butter	8%	52

餡料	Filling
馬來咖吔適量	kaya (coconut jam)

做法
method

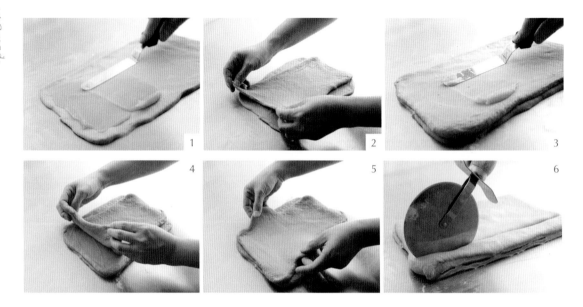

預備

• 將中種切成小塊

• 攪拌器放入斑蘭葉和水，攪爛，隔出汁液待用

混合： 將麵糰所有材料(牛油除外)混合，搓揉，逐小加入中種，搓至柔
　　　 滑，加入牛油再搓至可伸延薄膜。

發酵： 麵糰放入大碗內，蓋上保鮮紙，進行第一次發酵，約25-30分
　　　 鐘。

分割： 將麵糰分成三等份，排氣，按扁，輕捲成條狀，蓋上保鮮紙，靜
　　　 置20分鐘讓麵糰鬆弛。

造型： 將麵糰排氣，擀薄成長方形的麵皮，塗上馬來咖吔後，蓋上另一
　　　 塊麵皮，再塗上一層馬來咖吔，再蓋上另一層麵皮，成為三層
　　　 厚的麵糰；將麵糰切成約2厘米寬的長條，將每條長條扭成麻花
　　　 狀，再成繩結狀，入模，用保鮮紙蓋好(看圖1-10)。

最後發酵：麵糰發酵約30分鐘至兩倍大(看圖11)，塗上蛋液，放入已預熱
　　　 170-180℃的焗爐，焗約25-30分鐘至金黃。

Preparation

- Cut the pre-ferment dough into small pieces.
- Blend the pandan leaf in water. Strain the juice.

1. Knead all ingredients of dough together (except the butter). Add pre-ferment dough piece by piece. Knead after each addition until soft and smooth. Add butter. Knead until stretchable consistency.

2. Put the dough into a big bowl. Cover with a cling wrap. Let it prove for about 25-30 minutes.

3. Divide the dough into three equal portions. Flatten the dough with your hands to drive the air out and hand square each piece. Cover in cling wrap. Set aside to rest for about 20 minutes.

4. Flatten the dough with your hands to drive the air out. Roll each piece of dough out into a rectangle with a rolling pin. Spread a layer of kaya on one of them. Put another sheet over it. Spread another layer of kaya. Then cover with last sheet of dough. Cut the stack into 2-cm-wide long strips. Twist each strip and then tied into a knot. Place into a greased mould. Cover with cling wrap (see pictures 1-10).

5. Let it prove for about 30 minutes or until it doubles in size (see picture 11). Brush egg wash on top. Bake in a pre-heated oven at 170-180°C for about 25-30 minutes until golden brown.

tips

- 可用斑蘭糖代替砂糖和斑蘭汁，但味道稍遜。
- 用斑蘭葉重量的一倍半水分，就可榨出濃淡得宜的汁液。
- You may use pandan sugar instead of pandan juice and sugar. But it wouldn't taste as strong.
- To make pandan juice of the right concentration, always add 1.5 parts of water to 1 part of pandan leaves (by weight).

7

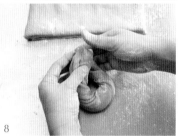

8

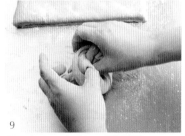

9

10

11

咖啡麵包
Coffee Bread

中種 Pre-ferment Dough	百分比%	克 gram
筋粉 bread flour	70%	469
水 water	40%	268
鮮酵母 fresh yeast	2%	13
海鹽 sea salt	1%	7
脫脂奶粉 skim milk powder	2%	13

酵母溶於水中，再加入筋粉、海鹽、脫脂奶粉搓至柔滑，用保鮮紙包裹麵糰，放入雪櫃發酵17小時。

Dissolve the yeast in water. Add bread flour, sea salt, skim milk powder and knead until soft. Cover the dough in cling wrap. Refrigerate to let it prove for 17 hours.

麵糰 Dough	百分比%	克 gram
筋粉 bread flour	30%	201
特濃咖啡粉 espresso coffee powder	2%	13
海鹽 sea salt	1%	7
砂糖 sugar	10%	67
蜂蜜 honey	8%	54
牛奶 milk	8%	54
蛋 egg	12%	80
鮮酵母 fresh yeast	1%	7
無鹽牛油 unsalted butter	10%	67

做法
method

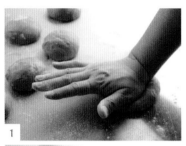

預備

- 將中種切成小塊

混合： 將麵糰所有材料(牛油除外)混合，搓揉，逐少加入中種，搓至柔滑，加入牛油再搓至可伸延薄膜。

發酵： 麵糰放入大碗內，蓋上保鮮紙，進行第一次發酵，約25-30分鐘。

分割： 麵糰分割成十六等份，排氣，滾圓，蓋上保鮮紙，靜置20分鐘讓麵糰鬆弛。

造型： 將麵糰排氣，滾成圓形，每8個麵糰放入吐司模內，輕輕蓋上保鮮紙(看圖1-4)。

最後發酵： 麵糰發酵約40-45分鐘至八成滿(看圖5)，塗上蛋液，放入已預熱170-180℃焗爐，焗約30-35分鐘至金黃。

貼士

- 我喜歡使用即溶特濃咖啡粉，味道和濃度絕佳；若買不到，可用慣用的咖啡粉或即煮咖啡，濃淡要自己調配。

Preparation

- Cut the pre-ferment dough into small pieces.

1. Knead all ingredients of dough together (except the butter). Add pre-ferment dough piece by piece. Knead after each addition until soft and smooth. Add butter. Knead until stretchable consistency.
2. Put the dough into a big bowl. Cover with a cling wrap. Let it prove for about 25-30 minutes.
3. Divide the dough into sixteen equal portions. Flatten each piece of dough with your hands to drive the air out and then round it. Cover in cling wrap. Set aside to rest for about 20 minutes.
4. Flatten the dough with your hands to drive the air out. Round each piece again. Put eight pieces of dough into the loaf tin and cover with cling wrap (see pictures 1-4).
5. Let it prove again for about 40-45 minutes or until the dough has risen to 80% of the depth of the mould (see picture 5). Brush egg wash on top. Bake in a pre-heated oven at 170-180°C for about 30-35 minutes until golden brown.

Tips

- I recommend using instant espresso powder for this bread because of its rich flavour. If you can't get it, you may use any instant coffee or freshly brewed coffee instead. Yet, you may have to adjust the strength or the amount accordingly.

伯爵奶茶麵包

Earl Grey Milk Tea Bread

材料 ingredients

中種 Pre-ferment Dough	百分比%	克 gram
筋粉 bread flour	70%	464
水 water	40%	265
鮮酵母 fresh yeast	2%	13
海鹽 sea salt	1%	7
脫脂奶粉 skim milk powder	2%	13

酵母溶於水中，再加入筋粉、海鹽、脫脂奶粉搓至柔滑，用保鮮紙包裹麵糰，放入雪櫃發酵17小時。

Dissolve the yeast in water. Add bread flour, sea salt, skim milk powder and knead until soft. Cover the dough in cling wrap. Refrigerate to let it prove for 17 hours.

麵糰 Dough	百分比%	克 gram
筋粉 bread flour	30%	199
脫脂奶粉 skim milk powder	4%	27
海鹽 sea salt	1%	7
砂糖 sugar	12%	80
蛋 egg	12%	80
濃縮奶茶 milk tea concentrate	12%	80
鮮酵母 fresh yeast	1%	7
無鹽牛油 unsalted butter	10%	66
伯爵茶葉碎 ground Earl Grey tea leaves	2%	13

濃縮奶茶	Milk tea concentrate
牛奶133克	133 g milk
伯爵茶葉 20克	20 g Earl Grey tea leaves

做法
method

預備

- 先預備濃縮奶茶：將牛奶煮熱，加入伯爵茶葉煮滾後關火，泡浸約15分鐘釋出香味。隔去茶葉後，量出所需份量

- 將中種切成小塊

混合：　將麵糰所有材料(牛油除外)混合，搓揉，逐少加入中種，搓至柔滑，加入牛油，再搓至可伸延薄膜。

發酵：　麵糰放入大碗內，蓋上保鮮紙，進行第一次發酵，約25-30分鐘。

分割：　將麵糰分成六等份，排氣。每份麵糰輕捲成條狀，蓋上保鮮紙，靜置20分鐘讓麵糰鬆弛。

造型：　將麵糰排氣，擀薄成長方形，兩邊向內摺起約2厘米，再擀薄，寬度與吐司模相若。捲起後放入吐司模，用保鮮紙蓋好(看圖1-7)。

最後發酵：麵糰發酵約40-45分鐘至八成滿(看圖8)，塗上蛋液，關上模蓋，放入已預熱170-180℃焗爐，焗約30-35分鐘至金黃。

貼士

- 我較喜歡用伯爵茶或仕女伯爵茶，它所含的佛手柑味是其他茶沒法比擬的。

Preparation

- To make the Milk Tea Concentrate, heat the milk and put in the tea leaves. Leave it to brew for 15 minutes for thorough infusion. Strain and take the amount required in the recipe.
- Cut the pre-ferment dough into small pieces.

1. Knead all ingredients of dough together (except the butter). Add pre-ferment dough piece by piece. Knead after each addition until soft and smooth. Add butter. Knead until stretchable consistency.

2. Put the dough into a big bowl. Cover with cling wrap and let it prove for about 25-30 minutes.

3. Divide the dough into six small equal portions. Flatten each portion with your hands to drive the air out. Hand square it and cover with cling wrap. Set aside to rest for about 20 minutes.

4. Flatten each piece of dough with your hands to drive the air out. Roll each out into a rectangular sheet with a rolling pin. Fold about 2 cm from both sides towards the centre and roll it out again. Roll the dough up (the width should be same as the loaf tin) and put them into the mould. Cover with cling wrap (see pictures 1-7).

5. Let it prove for about 40-45 minutes or until the dough has risen to 80% of the depth of the loaf tin (see picture 8). Close the lid. Bake in a pre-heated oven at 170-180°C for about 30-35 minutes.

Tips

- I recommend using Earl Grey or Lady Grey tea in this bread, as their bergamot fragrance is simply beyond compare.

5

6

7

8

芝士薯茸蘑菇洋蔥包

Mushroom Buns with Cheese and Mashed Potato

材料 ingredients

中種 Pre-ferment Dough	百分比%	克 gram
筋粉 bread flour	70%	381
水 water	40%	218
鮮酵母 fresh yeast	2%	11
海鹽 sea salt	1%	5
脫脂奶粉 skim milk powder	2%	11

酵母溶於水中，再加入筋粉、海鹽、脫脂奶粉搓至柔滑，用保鮮紙包裹麵糰，放入雪櫃發酵17小時。

Dissolve the yeast in water. Add bread flour, sea salt, skim milk powder and knead until soft. Cover the dough in cling wrap. Refrigerate to let it prove for 17 hours.

麵糰 Dough	百分比%	克 gram
筋粉 bread flour	30%	163
海鹽 sea salt	1.2%	7
砂糖 sugar	8%	44
水 water	16%	87
薯茸 mashed potato	15.5%	84
鮮酵母 fresh yeast	1%	5
橄欖油 olive oil	8%	44
炒香蘑菇 pan-fried button mushrooms	12.5%	68
炒香洋蔥 pan-fried onion	23.5%	128
大孔芝士 emmental cheese	12%	65
乾雜香草 mixed herbs		適量
黑椒 black pepper		適量

預備 *preparation*

- 將馬鈴薯連皮焓熟,去皮,壓成薯茸,待涼備用。洋葱切絲,蘑菇切片,用橄欖油炒香,隔去水分,待涼備用。大孔芝士切粒。
- 將中種切成小塊

- Boil the potatoes in water. Peel and mash it. Slice the onion and button mushrooms. Sauté in olive oil. Drain. Diced the emmental cheese.
- Cut the pre-ferment dough into small pieces.

做法 *method*

混合:　將麵糰所有材料(洋葱、大孔芝士、蘑菇除外)混合,搓揉,逐少加入中種,搓至柔滑,加入洋葱、大孔芝士、蘑菇,搓至可伸延薄膜。

發酵:　麵糰放入大碗內,蓋上保鮮紙進行第一次發酵,約25-30分鐘。

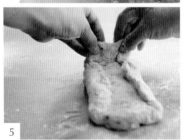

分割:　將麵糰分成六等份,排氣,按扁,輕捲成條狀,蓋上保鮮紙,靜置20分鐘讓麵糰鬆弛。

造型:　將麵糰排氣,擀薄成長方形,兩邊向內摺入約2厘米,捲起,收緊,放模內發酵(看圖1-10),蓋上保鮮紙。

最後發酵:麵糰發酵45分鐘或至兩倍大,塗上蛋液(看圖11),放入已預熱170-180℃焗爐,焗約30-35分鐘至金黃。

1. Knead all ingredients of dough together (except the onion, emmental cheese and button mushrooms). Add pre-ferment dough piece by piece. Knead after each addition until soft and smooth. Add onion, emmental cheese and button mushrooms. Knead until stretchable consistency.

2. Put the dough into a big bowl. Cover with a cling wrap. Let it prove for about 25-30 minutes.

3. Divide the dough into six equal portions. Flatten each piece with your hands to drive the air out and hand square them. Cover with cling wrap. Set aside to rest for about 20 minutes.

4. Flatten each piece of dough with your hands to drive the air out. Roll the dough out with a rolling pin into a rectangle. Fold about 2 cm from both sides towards the centre. Roll each piece up. Put a piece of dough into a baking mould (see pictures 1-10). Cover with cling wrap.

5. Let it prove for about 45 minutes or until the dough has doubled in size. Brush egg wash on top (see picture 11). Bake in a pre-heated oven at 170-180°C for about 30-35 minutes until golden brown.

tips

- 宜選用高澱粉的馬鈴薯；大孔芝士比較不易溶，同時較標榜高溶點的芝士健康。

- 不銹鋼模具要塗軟牛油，會較易倒出已烘焙的麵包。

- Use russet potatoes for their high starch content. Emmental cheese stands heat well and is considered healthier than processed cheddar cheese.

- It's better to grease stainless steel mould with soft butter for this bun as you can unmould it more easily.

雜穀麥麵包
Multi-grain Bread

材料 ingredients

中種 Pre-ferment Dough	百分比%	克 gram
筋粉 bread flour	70%	484
水 water	40%	276
鮮酵母 fresh yeast	2%	14
海鹽 sea salt	1%	7
脫脂奶粉 skim milk powder	2%	14

酵母溶於水中，再加入筋粉、海鹽、脫脂奶粉搓至柔滑，用保鮮紙包裹麵糰，放入雪櫃發酵17小時。

Dissolve the yeast in water. Add bread flour, sea salt, skim milk powder and knead until soft. Cover the dough in cling wrap. Refrigerate to let it prove for 17 hours.

麵糰 Dough	百分比%	克 gram
筋粉 bread flour	5%	35
黑麥粉 rye flour	10%	69
麥芽精 malt extract	2%	14
全麥粉 whole wheat flour	15%	104
砂糖 sugar	6%	41
海鹽 sea salt	1%	7
水 water	29%	200
鮮酵母 fresh yeast	1%	7
橄欖油 olive oil	7%	48
葵花籽 sunflower seeds	7.7%	40
亞麻籽 flaxseeds	9.6%	50
燕麥粒 oat groats	29%	150
小麥粒 wheat	29%	150

做法
method

預備

- 將葵花籽烘香;亞麻籽浸軟,用水洗去膠質;燕麥粒、小麥粒用水煮約半小時至軟身,待涼備用

- 將中種切成小塊

混合:	將麵糰所有材料(雜穀麥除外)混合,搓揉,逐少加入中種,搓至柔滑,加入雜穀麥再搓至可伸延薄膜。
發酵:	麵糰放入大碗內,蓋上保鮮紙進行第一次發酵,約25-30分鐘。
分割:	麵糰分割成八等份,排氣,滾圓,靜置20分鐘讓麵糰鬆弛。
造型:	在籐製麵包籃上均勻地篩上筋粉。將麵糰排氣,滾圓,收口向上,四個一組,將麵糰放入籃內發酵,輕輕蓋上保鮮紙(看圖1-8)。
最後發酵:	麵糰發酵約40分鐘至兩倍大,把發酵籃內的麵糰輕輕倒轉在焗盤上,放入已預熱170-180℃的焗爐內,在爐內噴水製造蒸汽效果,焗約30-35分鐘至金黃(看圖9-11)。

貼士

- 如沒有發酵籃,可用筲箕和薄布代替。只需將布放在筲箕上,再篩上筋粉即成(看圖12)。
- 麥芽精在各大超級市場和烘焙用品店有售。

Preparation

- Briefly bake the sunflower seeds. Soak the flaxseeds in water and drain. Wash them repeatedly to remove the gluey slime. Boil oat groats and wheat in water for about half an hour until soft. Drain and leave them to cool.

- Cut the pre-ferment dough into small pieces.

1. Knead all ingredients of dough together (except the seeds and grains). Add pre-ferment dough piece by piece. Knead after each addition until soft and smooth. Add the seeds and grains. Knead until stretchable consistency.

2. Put the dough into a big bowl. Cover with a cling wrap. Let it prove for about 25-30 minutes.

3. Divide the dough into eight equal portions. Flatten each piece of dough with your hands to drive the air out and then round them. Set aside to rest for about 20 minutes.

4. Sift bread flour evenly on the proving baskets. Flatten each piece of dough with your hands to drive the air out. Round them. Put four pieces of dough into a floured proving basket (you should put the seam side up). Cover with cling wrap (see pictures 1-8).

5. Let it prove again for about 40 minutes or until they double in size. Unmould the dough from the proving basket by turning them upside down to the baking tray. Bake in a pre-heated oven at 170-180°C for about 30-35 minutes. Spray water in the oven to create a steamy atmosphere (see pictures 9-11).

Tips

- If you don't have a proving basket, you can always replace it with a strainer and a piece of cheese cloth. Just line the strainer with the cheese cloth. Sprinkle bread flour all over. (see picture 12)

- You may get malt extract in major supermarkets and baking supply stores.

楓糖栗子麵包
Chestnut and Maple Syrup Bread

材料
ingredients

中種 Pre-ferment Dough	百分比%	克 gram
筋粉 bread flour	70%	440
水 water	40%	251
鮮酵母 fresh yeast	2%	13
海鹽 sea salt	1%	6
脫脂奶粉 skim milk powder	2%	13

酵母溶於水中，再加入筋粉、海鹽、脫脂奶粉搓至柔滑，用保鮮紙包裹麵糰，放入雪櫃發酵17小時。

Dissolve the yeast in water. Add bread flour, sea salt, skim milk powder and knead until soft. Cover the dough in cling wrap. Refrigerate to let it prove for 17 hours.

麵糰 Dough	百分比%	克 gram
筋粉 bread flour	30%	189
脫脂奶粉 skim milk powder	3%	19
海鹽 sea salt	1%	6
楓糖漿 maple syrup	16%	101
砂糖 sugar	3%	19
蛋 egg	8%	50
鮮酵母 fresh yeast	1%	6
無鹽牛油 unsalted butter	8%	50
罐頭栗子茸 canned chestnut puree	25%	157
熟栗子粒 cooked chestnuts		適量

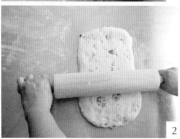

做法
method

預備

- 將中種切成小塊

混合： 將麵糰所有材料(牛油、熟栗子粒除外)混合，搓揉，逐少加入中種，搓至柔滑，加入牛油、熟栗子粒，搓至可伸延薄膜。

發酵： 麵糰放入大碗內，蓋上保鮮紙，進行第一次發酵，約25-30分鐘。

分割： 將麵糰分成兩等份，排氣。每份麵糰輕捲成條狀，蓋上保鮮紙，靜置20分鐘讓麵糰鬆弛。

造型： 將麵糰排氣，擀薄成長方形，兩邊向內摺起約4厘米，再擀薄，寬度與吐司模相若。捲起後放入吐司模，用保鮮紙蓋好(看圖1-8)。

最後發酵： 麵糰發酵約45分鐘至八成滿，在麵糰上噴水，篩上筋粉，劃上花紋，放入已預熱170-180℃焗爐，焗約30-35分鐘至金黃(看圖9-12)。

貼士

- 楓糖漿是加拿大的產品，但很奇怪，朋友從加拿大帶回來的糖漿遠不及在香港買的香醇。同時，要注意不要買錯楓糖味糖漿啊！
- 可用罐裝或急凍熟栗子。

7

Preparation

- Cut the pre-ferment dough into small pieces.

1. Knead all ingredients of dough together (except the butter and cooked chestnuts). Add pre-ferment dough piece by piece. Knead after each addition until soft and smooth. Add butter and cooked chestnuts. Knead until stretchable consistency.

2. Put the dough into a big bowl. Cover with a cling wrap. Let it prove for about 25-30 minutes.

3. Divide the dough into two equal portions. Flatten the dough with your hands to drive the air out and hand square them. Cover with cling wrap. Set aside to rest for about 20 minutes.

4. Flatten the dough with your hand to drive the air out. Roll each piece of dough out into a rectangle. Fold about 4 cm on each side towards the centre. Roll the dough out with rolling pin again into a rectangle (about the same width as the baking mould). Place the dough into the mould. Cover with cling wrap (see pictures 1-8).

5. Let it prove for about 45 minutes or until the dough has risen to 80% of the depth of the mould. Spray water on top and sift bread flour on it. Score couple times on top along the length. Bake in a pre-heated oven at 170-180°C for about 30-35 minutes until golden brown (see pictures 9-12).

8

9

10

Tips

- Maple syrup is the sweet sap extracted from maple trees and is a Canadian specialty. Strangely enough, the maple syrup my friends brought me from Canada never tastes as rich as those I get in Hong Kong. Besides, pay attention to the label. Get genuine maple syrup, not maple-flavoured syrup.

- You can use canned cooked chestnuts or frozen cooked chestnuts.

11

12

豆腐麥麩麵包

Tofu and Wheat Bran Bread

材料 ingredients

中種 Pre-ferment Dough	百分比%	克 gram
筋粉 bread flour	70%	422
水 water	40%	241
鮮酵母 fresh yeast	2%	12
海鹽 sea salt	1%	6
脫脂奶粉 skim milk powder	2%	12

酵母溶於水中，再加入筋粉、海鹽、脫脂奶粉搓至柔滑，用保鮮紙包裹麵糰，放入雪櫃發酵17小時。

Dissolve the yeast in water. Add bread flour, sea salt, skim milk powder and knead until soft. Cover the dough in cling wrap. Refrigerate to let it prove for 17 hours.

麵糰 Dough	百分比%	克 gram
筋粉 bread flour	30%	181
麥麩 wheat bran	4%	24
即沖無糖豆漿粉 unsweetened soybean milk powder	8%	48
海鹽 sea salt	1%	6
砂糖 sugar	12%	72
濃味豆腐 beancurd	40%	241
鮮酵母 fresh yeast	0.8%	5
無鹽牛油 unsalted butter	8%	48

做法
method

預備

- 將中種切成小塊
- 擠去豆腐水分,待用

混合: 將麵糰所有材料(牛油除外)混合,搓揉,逐少加入中種,搓至柔滑,加入牛油再搓至可伸延薄膜。

發酵: 麵糰放入大碗內,蓋上保鮮紙進行第一次發酵,約25-30分鐘。

分割: 將麵糰分成六等份,排氣,滾圓,靜置20分鐘讓麵糰鬆弛。

造型: 籐製麵包籃內均勻篩上筋粉。將麵糰排氣,滾圓,三個一組,收口向上,將麵糰放在籃內發酵,輕輕蓋上保鮮紙(看圖1-6)。

最後發酵: 麵糰發酵約40分鐘至兩倍大,把發酵籃內的麵糰輕輕倒落焗盤上。放入已預熱170-180℃的焗爐內,在爐內噴水製造蒸汽效果,焗約35-40分鐘至金黃(看圖7-8)。

貼士

- 豆腐會出水,須注意水分

Preparation

- Cut the pre-ferment dough into small pieces.
- Press the beancurd to drain the water.

1. Knead all ingredients of dough together (except the butter). Add pre-ferment dough piece by piece. Knead after each addition until soft and smooth. Add butter. Knead until stretchable consistency.
2. Put the dough into a big bowl. Cover with cling wrap and let it prove for about 25-30 minutes.
3. Divide the dough into six equal portions. Flatten each piece of dough with your hands to drive the air out. Round each of them. Set aside to rest for about 20 minutes
4. Sift bread flour evenly on the proving baskets. Flatten each piece of dough to drive the air of it. Round each piece again. Place three pieces of dough into one floured proving basket (you should put the seam side up). Cover with cling wrap (see pictures 1-6).
5. Let it prove again for about 40 minutes or until the dough doubles in size. Unmould and put them on a baking tray. Pre-heat an oven at 170-180°C. Spray water in the oven to create a steamy atmosphere. Bake for about 35-40 minutes (see pictures 7-8).

Tips

- Tofu tends to release much water and you should pay attention to the water content of the dough.

五穀肉鬆麵包
Multi-grain Rolls with Pork Floss

中種 Pre-ferment Dough	百分比%	克 gram
筋粉 bread flour	70%	438
水 water	40%	250
鮮酵母 fresh yeast	2%	13
海鹽 sea salt	1%	6
脫脂奶粉 skim milk powder	2%	13

酵母溶於水中，再加入筋粉、海鹽、脫脂奶粉搓至柔滑，用保鮮紙包裹麵糰，放入雪櫃發酵17小時。

Dissolve the yeast in water. Add bread flour, sea salt, skim milk powder and knead until soft. Cover the dough in cling wrap. Refrigerate to let it prove for 17 hours.

麵糰 Dough	百分比%	克 gram
筋粉 bread flour	30%	188
脫脂奶粉 skim milk powder	3%	19
五穀粉 ground multi-grains	15%	94
海鹽 sea salt	1%	6
砂糖 sugar	10%	63
水 water	29%	181
鮮酵母 fresh yeast	1%	6
無鹽牛油 unsalted butter	7%	44
沙律醬 creamy salad dressing		適量
豬肉鬆 dried pork floss		適量

做法
method

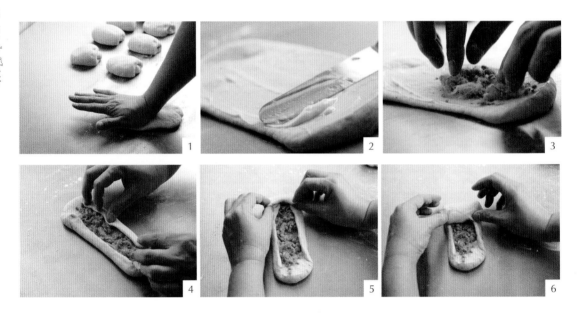

預備

● 將中種切成小塊

混合： 將麵糰所有材料(牛油、沙律醬、豬肉鬆除外)混合，搓揉，逐少
加入中種，搓至柔滑，加入牛油再搓至可伸延薄膜。

發酵： 麵糰放入大碗內，蓋上保鮮紙進行第一次發酵，約25-30分鐘。

分割： 將麵糰分成十六等份，排氣，按扁，輕捲成條狀，蓋上保鮮紙，
靜置20分鐘讓麵糰鬆弛。

造型： 將麵糰排氣，擀薄成長方形，塗沙律醬，灑上豬肉鬆，左右兩邊
摺入約2厘米，捲起；每8卷麵糰放模內發酵，蓋上保鮮紙(看圖
1-10)。

最後發酵： 麵糰發酵約45分鐘至八成滿(看圖11)，塗上蛋液，放入已預熱
170-180℃的焗爐，焗約30-35分鐘至金黃。

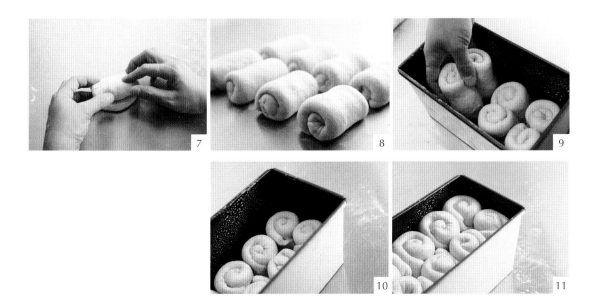

Preparation

- Cut the pre-ferment dough into small pieces.

1. Knead all ingredients of dough together (except the butter, dried pork floss and the salad dressing). Add pre-ferment dough piece by piece. Knead after each addition until soft and smooth. Add butter. Knead until stretchable consistency.
2. Put the dough into a big bowl. Cover with cling wrap and let it prove for about 25-30 minutes.
3. Divide the dough into sixteen equal portions. Flatten each portion with your hands to drive the air out. Then hand square each portion and cover with cling wrap. Set aside to rest for about 20 minutes.
4. Flatten each piece of dough with your hands to drive the air out. Roll each into a rectangle with a rolling pin. Spread salad dressing and sprinkle dried pork floss on top. Fold about 2 cm of dough from both edges towards the centre. Roll it up tightly. Place eight pieces of dough into a loaf tin. Cover with cling wrap (see pictures 1-10).
5. Let it prove for about 45 minutes or until the dough has risen to 80% of the depth of the mould (see picture 11). Brush egg wash on the surface. Bake in a pre-heated oven at 170-180°C for about 30-35 minutes.

南瓜葵花籽包
Pumpkin and Sunflower Seed Loaf

中種 Pre-ferment Dough	百分比%	克 gram
筋粉 bread flour	70%	411
水 water	40%	235
鮮酵母 fresh yeast	2%	12
海鹽 sea salt	1%	6
脫脂奶粉 skim milk powder	2%	12

酵母溶於水中，再加入筋粉、海鹽、脫脂奶粉搓至柔滑，用保鮮紙包裹麵糰，放入雪櫃發酵17小時。

Dissolve the yeast in water. Add bread flour, sea salt, skim milk powder and knead until soft. Cover the dough in cling wrap. Refrigerate to let it prove for 17 hours.

麵糰 Dough	百分比%	克 gram
筋粉 bread flour	30%	176
脫脂奶粉 skim milk powder	3%	18
海鹽 sea salt	1%	6
砂糖 sugar	10%	59
蛋 egg	3%	18
鮮酵母 fresh yeast	1%	6
水 water	7%	41
無鹽牛油 unsalted butter	7%	41
熟南瓜茸 cooked pumpkin puree	23%	135
生南瓜絲 julienne pumpkin	15%	88
烘香葵花籽 baked sunflower seeds	10%	59

裝飾	Garnish
葵花籽適量	sunflower seeds

預備 *preparation*

- 南瓜一個,一半連皮蒸熟,起肉,壓成茸;另一半南瓜刨成幼絲

- 葵花籽放焗爐稍為烘香,待涼備用

- 將中種切成小塊

- Steam half the pumpkin with skin on until tender. Then peel, seed and mash. Julienne the remaining half of the pumpkin.

- Bake the sunflower seeds in an oven briefly. Leave them to cool.

- Cut the pre-ferment dough into small pieces.

做法 *method*

混合: 將麵糰所有材料(牛油、生南瓜絲、烘香葵花籽除外)混合,搓揉,逐少加入中種搓至柔滑,加入牛油、生南瓜絲、烘香葵花籽再搓至可伸延薄膜。

發酵: 麵糰放入大碗內,蓋上保鮮紙進行第一次發酵,約25-30分鐘。

分割: 將麵糰分成兩等份,排氣。每份麵糰輕捲成條狀,靜置20分鐘讓麵糰鬆弛。

造型: 將麵糰排氣,擀薄成長方形,兩邊向內摺起,再擀薄,捲成跟吐司模長度相若的卷狀。在麵糰上噴水,滾上裝飾用葵花籽,放入吐司模,用保鮮紙蓋好(看圖1-10)。

最後發酵:麵糰發酵約45分鐘至八成滿(看圖11),放入已預熱170-180℃的焗爐內,焗約30-35分鐘至金黃。

1. Knead all ingredients together (except the butter, sunflower seeds and the pumpkin julienne). Add pre-ferment dough piece by piece. Knead after each addition until soft and smooth. Add butter, sunflower seeds and the pumpkin julienne. Knead until stretchable consistency.

2. Put the dough into a big bowl. Cover with cling wrap and let it prove for about 25-30 minutes.

3. Divide the dough into two equal parts and hand square each. Set aside to rest for about 20 minutes.

4. Flatten the dough with your hands to drive the air out. Roll the dough out with a rolling pin into a rectangle. Fold both ends inwards and then roll it out once more. Roll the dough up to about the same width of the loaf tin. Spray water on it and coat the dough in sunflower seeds. Place dough into a loaf tin. Cover with cling wrap (see pictures 1-10).

5. Let it prove for about 45 minutes or until the dough has risen to 80% of the depth of the mould (see picture 11). Bake in a pre-heated oven at 170-180°C for about 30-35 minutes.

貼士
tips

- 本書使用的是日本南瓜,它味道甘香甜美,水分適中,是優質的食材。如選用其他南瓜如本地南瓜,水分較多,蒸熟後要隔去水分才可使用,至於泰國南瓜或大陸出產的日本種南瓜,甜味不足而水分少,做時須酌量調校水分。

- The recipe here uses premium Japanese pumpkin. Local pumpkin is cheaper but they are not as sweet and tend to release more water after cooked. If you use local pumpkin, make sure you drain the water well after steaming it. Thai pumpkins or those Japanese pumpkins grown in mainland China are not sweet enough and their water content is too low. You may have to adjust the amount of water used in the dough.

合桃無花果麵包

Walnut and Fig Bread

材料 ingredients

中種 Pre-ferment Dough	百分比%	克 gram
筋粉 bread flour	70%	373
水 water	40%	213
鮮酵母 fresh yeast	2%	11
海鹽 sea salt	1%	5
脫脂奶粉 skim milk powder	2%	11

酵母溶於水中，再加入筋粉、海鹽、脫脂奶粉搓至柔滑，用保鮮紙包裹麵糰，放入雪櫃發酵17小時。

Dissolve the yeast in water. Add bread flour, sea salt, skim milk powder and knead until soft. Cover the dough in cling wrap. Refrigerate to let it prove for 17 hours.

麵糰 Dough	百分比%	克 gram
筋粉 bread flour	20%	106
黑麥粉 rye flour	12%	64
海鹽 sea salt	1%	5
砂糖 sugar	15%	80
水 water	27%	144
鮮酵母 fresh yeast	1%	5
無鹽牛油 unsalted butter	7%	37
乾無花果 dried fig	25%	133
合桃 walnut	25%	133

預備 *preparation*

- 將中種切成小塊
- 合桃稍微烘香，然後和無花果一起切成小塊
- Cut the pre-ferment dough into small pieces.
- Bake the walnuts briefly. Chop the dried figs and walnuts into small pieces.

混合：　　將麵糰所有材料(牛油、無花果和合桃除外)混合，搓揉，逐少加入中種，搓至柔滑，加入牛油、無花果和合桃搓至可伸延薄膜。

發酵：　　麵糰放入大碗內，蓋上保鮮紙，進行第一次發酵，約25-30分鐘。

分割：　　將麵糰分成六等份，排氣，滾圓，蓋上保鮮紙，靜置20分鐘讓麵糰鬆弛。

造型：　　將麵糰排氣，捲成橄核形，將一端搓尖使麵糰成葉尖狀，放在已灑粉的帆布上發酵，輕輕蓋上保鮮紙(看圖1-3)。

最後發酵：麵糰發酵約40-45分鐘至兩倍大，噴上少許水分，篩上黑麥粉，用刀片劃上花紋(每邊約7-8刀)，放入已預熱170-180℃焗爐，焗約25-30分鐘至金黃，初期要在焗爐中噴水製造蒸氣效果(看圖4-8)。

1. Knead all ingredients of dough together (except the butter, walnuts and dried figs). Add pre-ferment dough piece by piece. Knead after each addition until soft and smooth. Add butter, walnuts and dried figs. Knead until stretchable consistency.

2. Put the dough into a big bowl. Cover with cling wrap and let it prove for about 25-30 minutes.

3. Divide the dough into six equal portions. Flatten each portion with your hands to drive the air out. Round it. Cover with cling wrap and set aside to rest for about 20 minutes.

4. Flatten the dough with your hands to drive the air out. Knead each piece of dough into an oval shape. Roll one end to make it pointy. Put them on floured canvas. Cover with cling wrap (see pictures 1-3).

5. Let it prove for about 40-45 minutes or until they double in size. Spray water over them. Sieve rye flour on top. Score it to create veins on a leaf (7 to 8 scores each side). Bake in a pre-heated oven at 170-180°C for about 25-30 minutes (spray water inside the oven at the beginning to create steam in the baking process) (see pictures 4-8).

貼士
tips

- 請盡量選用優質無花果和合桃
- Use premium figs and walnuts if you can.

黑白芝麻麵包

Black and White Sesame Mini Loaf

材料 ingredients

中種 Pre-ferment Dough	百分比%	克 gram
筋粉 bread flour	70%	449
水 water	40%	256
鮮酵母 fresh yeast	2%	13
海鹽 sea salt	1%	6
脫脂奶粉 skim milk powder	2%	13

酵母溶於水中，再加入筋粉、海鹽、脫脂奶粉搓至柔滑，用保鮮紙包裹麵糰，放入雪櫃發酵17小時。

Dissolve the yeast in water. Add bread flour, sea salt, skim milk powder and knead until soft. Cover the dough in cling wrap. Refrigerate to let it prove for 17 hours.

麵糰 Dough	百分比%	克 gram
筋粉 bread flour	30%	192
脫脂奶粉 skim milk powder	3%	19
海鹽 sea salt	1%	6
砂糖 sugar	10%	64
水 water	21%	135
蛋 egg	9%	58
鮮酵母 fresh yeast	1%	6
無鹽牛油 unsalted butter	8%	51
黑芝麻粉 black sesame powder	8%	51
甜黑芝麻醬 sweeten black sesame paste		適量
白芝麻 white sesame		適量

做法
method

預備

● 將中種切成小塊

混合：　　將麵糰所有材料(牛油除外)混合，搓揉，逐少
　　　　　加入中種，搓至柔滑，加入牛油再搓至可伸延
　　　　　薄膜。

發酵：　　麵糰放入大碗內，蓋上保鮮紙，進行第一次發
　　　　　酵，約25-30分鐘。

分割：　　將麵糰分成六等份，排氣。輕捲成條狀，蓋上
　　　　　保鮮紙，靜置20分鐘讓麵糰鬆弛。

造型：　　將麵糰排氣，擀薄麵糰成長方形，塗上甜芝麻
　　　　　醬，左右兩邊向內覆入約2厘米，捲起(寬度與
　　　　　吐司模相若)(看圖1-7)。表面噴水，滾上白芝
　　　　　麻，放入吐司模內，蓋上保鮮紙(看圖8-9)。

最後發酵：麵糰發酵約45分鐘至2倍大(看圖10)，放入已預
　　　　　熱170-180℃焗爐，焗約30-35分鐘至金黃。

Preparation

- Cut the pre-ferment dough into small pieces.

1. Knead all ingredients of dough together (except the butter). Add pre-ferment dough piece by piece. Knead after each addition until soft and smooth. Add butter. Knead until stretchable consistency.

2. Put the dough into a big bowl. Cover with cling wrap and let it prove for about 25-30 minutes.

3. Divide the dough into six small equal portions. Flatten each portion with your hands to drive the air out and hand square them. Cover the dough in cling wrap. Set aside to rest for about 20 minutes.

4. Flatten the dough with your hands to drive the air out. Roll all six pieces of dough together into a rectangular sheet. Spread black sesame paste on top. Fold about 2 cm from both edges towards to centre. Then roll the dough up from one end (the width should be same as loaf tin), (see pictures 1-7). Spray water on the roll. Toss white sesames on top. Place the dough into a loaf tin. Cover with cling wrap (see pictures 8-9).

5. Let it prove for about 45 minutes or until the dough doubles in size (see picture 10). Bake in a pre-heated oven at 170-180°C for about 30-35 minutes.

6

7

8

9

10

麥麩藍芝士麵包

Wheat Bran
and Blue Cheese Bread

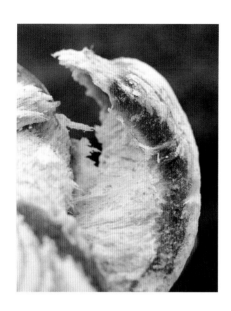

材料 ingredients

中種 Pre-ferment Dough	百分比%	克 gram
筋粉 bread flour	70%	456
水 water	40%	261
鮮酵母 fresh yeast	2%	13
海鹽 sea salt	1%	7
脫脂奶粉 skim milk powder	2%	13

酵母溶於水中，再加入筋粉、海鹽、脫脂奶粉搓至柔滑，用保鮮紙包
裹麵糰，放入雪櫃發酵17小時。

Dissolve the yeast in water. Add bread flour, sea salt, skim milk powder and
knead until soft. Cover the dough in cling wrap. Refrigerate to let it prove
for 17 hours.

麵糰 Dough	百分比%	克 gram
筋粉 bread flour	30%	196
麥麩 wheat bran	6%	39
海鹽 sea salt	0.5%	3
砂糖 sugar	8%	52
水 water	29%	189
鮮酵母 fresh yeast	1%	7
藍芝士 blue cheese	13%	85

預備
preparation

● 將中種切成小塊

● Cut the pre-ferment dough into small pieces.

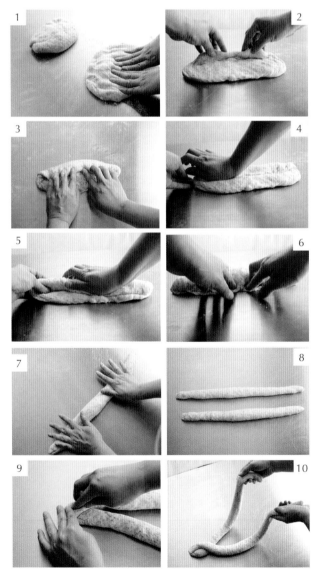

做法
method

混合：　　將麵糰所有材料混合，搓揉，逐少加入中種，搓至可伸延薄膜。

發酵：　　麵糰放入大碗內，蓋上保鮮紙進行第一次發酵，約25-30分鐘。

分割：　　將麵糰分成四等份，排氣。每份麵糰輕捲成條狀，靜置20分鐘讓麵糰鬆弛。

造型：　　在2個籐製麵包籃上均勻篩上筋粉。將麵糰排氣，再搓成條狀，兩條麵糰扭在一起；將麵糰頭尾摺疊，使麵糰成圓形，中間略留空間，放入籃內發酵，輕輕蓋上保鮮紙(看圖1-18)。

最後發酵：麵糰發酵約40分鐘至兩倍大，把發酵籃倒轉，麵糰輕輕倒在焗盤上，放入已預熱170-180℃的焗爐內，在爐內噴水製造蒸汽效果，焗約30分鐘至金黃(看圖19-20)。

1. Knead all ingredients of dough together. Add pre-ferment dough piece by piece. Knead after each addition until stretchable consistency.

2. Put the dough into a big bowl. Cover with a cling wrap. Let it prove for about 25-30 minutes.

3. Divide the dough into four equal portions and hand square each piece of dough. Set aside to rest for about 20 minutes.

4. Sift flour on two proving baskets evenly. Flatten each piece of dough with your hands to drive air out. Roll each with your hands into a long strand. Twist two strands together. Join both ends to form a circle. Put the dough into a floured proving basket. Cover with cling wrap (see pictures 1-18).

5. Let it prove for about 40 minutes or until the dough doubles in size. Unmould from the basket and place it on the baking tray. Bake in a pre-heated oven at 170-180°C for about 30 minutes. Spray water in the oven to create a steamy atmosphere in the baking process (see pictures 19-20).

tips

- 造型時，扭動得越緊，就會有更多圈紋，但切勿過度。
- 用藍芝士代替油脂；可以自己調整份量，但不宜太濃，可選用史堤頓藍芝士 (Stilton cheese)。
- The more you twist the two strands of dough together, the more spiral patterns there will be on the bread. But don't go too far.
- The blue cheese has much fat in it and that's why no grease is added to the dough. You may adjust the amount of blue cheese but don't add too much. I recommend using Stilton cheese for this bread.

紅酒紅莓山合桃麵包

Pecan Bread with Wine-poached Cranberries

材料 ingredients

中種 Pre-ferment Dough	百分比%	克 gram
筋粉 bread flour	70%	380
水 water	40%	217
鮮酵母 fresh yeast	2%	11
海鹽 sea salt	1%	5
脫脂奶粉 skim milk powder	2%	11

酵母溶於水中，再加入筋粉、海鹽、脫脂奶粉搓至柔滑，用保鮮紙包裹麵糰，放入雪櫃發酵17小時。

Dissolve the yeast in water. Add bread flour, sea salt, skim milk powder and knead until soft. Cover the dough in cling wrap. Refrigerate to let it prove for 17 hours.

麵糰 Dough	百分比%	克 gram
筋粉 bread flour	15%	81
黑麥 rye flour	15%	81
海鹽 sea salt	1%	5
砂糖 sugar	15%	81
水 water	18%	98
鮮酵母 fresh yeast	1%	5
無鹽牛油 unsalted butter	7%	38
紅莓乾 dried cranberries	25%	136
山合桃 pecans	25%	136
煮紅莓乾紅酒	6%	33
red wine reduction from cooking cranberries		
紅酒200克	200 g red wine	

做法
method

預備

- 紅莓乾用紅酒200克煮至紅酒收乾水份。紅莓乾、紅酒分開備用,此時紅酒約有33克。

- 烘香山合桃

- 將中種切成小塊

混合: 將麵糰所有材料(牛油、紅莓乾、山合桃除外)混合,搓揉,逐少加入中種,搓至柔滑,加入牛油、紅莓乾、山合桃再搓至可伸延薄膜。

發酵: 麵糰放入大碗內,蓋上保鮮紙進行第一次發酵,約25-30分鐘。

分割: 將麵糰分成六等份,排氣,滾圓,蓋上保鮮紙,靜置20分鐘讓麵糰鬆弛。

造型: 將麵糰排氣,捲成欖核形,放帆布上發酵(看圖1),輕輕蓋上保鮮紙或膠布。

最後發酵: 麵糰發酵約30分鐘至兩倍大,放焗盤上,噴水,篩上黑麥粉,在每個小包中間劃一刀,入爐,初期在爐內噴水製造蒸汽效果,焗約25-30分鐘至金黃(看圖2-4)。

Preparation

- Cook cranberries in red wine until the wine reduces to 33 g. Strain to separate the cranberries and red wine reduction. Leave them to cool.
- Bake the pecans briefly.
- Cut the pre-ferment dough into small pieces.

1. Knead all ingredients of dough together (except the butter, cranberries and pecans). Add pre-ferment dough piece by piece. Knead after each addition until soft and smooth. Add butter, cranberries and pecans. Knead until stretchable consistency.

2. Put the dough into a big bowl. Cover with cling wrap and let it prove for about 25-30 minutes.

3. Divide the dough into six equal portions. Flatten each piece of dough with your hands to drive the air out. Round them and cover in cling wrap. Set aside to rest for about 20 minutes.

4. Flatten the dough with your hands to drive the air out. Round them again and shape them like an oval. Put them on a canvas (see picture 1). Cover with cling wrap.

5. Let them prove for about 30 minutes or until they double in size. Spray water on the dough. Sift rye flour on top. Score them once along the length. Bake in a pre-heated oven at 170-180°C for about 25-30 minutes until golden brown. Spray water inside the oven to create a steamy atmosphere at the beginning of baking.

tips

- 使用最便宜的紅酒即可
- Just pick the cheapest red wine for this recipe.

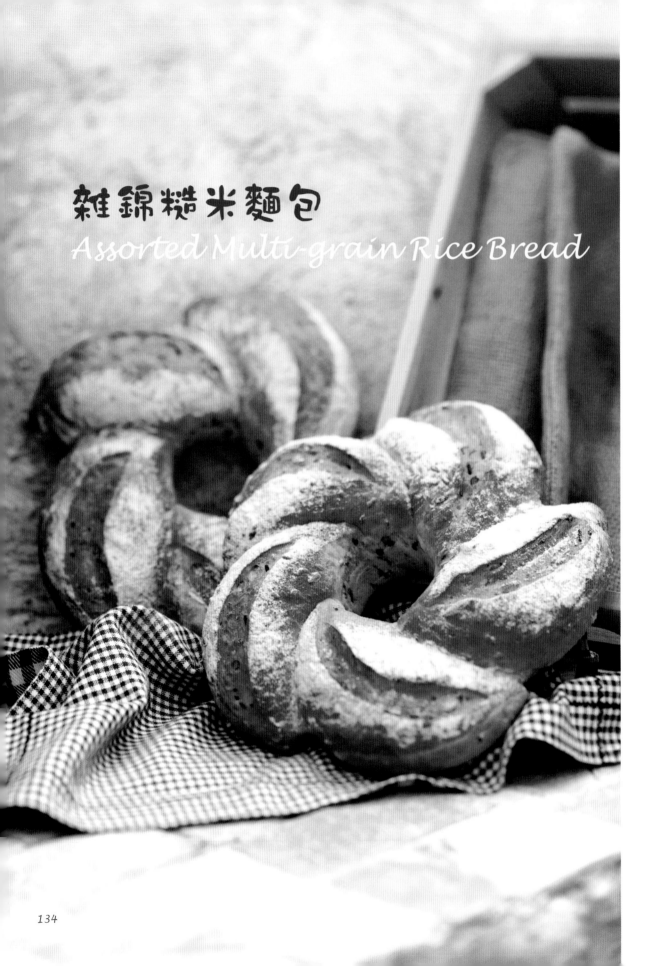

雜錦糙米麵包
Assorted Multi-grain Rice Bread

中種 Pre-ferment Dough	百分比%	克 gram
筋粉 bread flour	70%	404
水 water	40%	231
鮮酵母 fresh yeast	2%	12
海鹽 sea salt	1%	6
脫脂奶粉 skim milk powder	2%	12

酵母溶於水中，再加入筋粉、海鹽、脫脂奶粉搓至柔滑，用保鮮紙包裹麵糰，放入雪櫃發酵17小時。

Dissolve the yeast in water. Add bread flour, sea salt, skim milk powder and knead until soft. Cover the dough in cling wrap. Refrigerate to let it prove for 17 hours.

麵糰 Dough	百分比%	克 gram
筋粉 bread flour	30%	173
海鹽 sea salt	1%	6
砂糖 sugar	10%	58
水 water	25%	144
鮮酵母 fresh yeast	0.5%	3
無鹽牛油 unsalted butter	7%	40
糙米飯 cooked multi-grain rice	40%	231

預備
preparation

- 將雜錦糙米煮成飯，待涼備用
- 將中種切成小塊
- Cook multi-grain rice in a rice cooker until done. Set aside for later use.
- Cut the pre-ferment dough into small pieces.

做法
method

混合： 將麵糰所有材料(牛油、糙米飯除外)混合，搓揉，逐少加入中種，搓至柔滑，加入牛油、糙米飯再搓至可伸延薄膜。

發酵： 麵糰放入大碗內，蓋上保鮮紙進行第一次發酵，約25-30分鐘。

分割： 麵糰分割成十等份，排氣，滾圓，靜置20分鐘讓麵糰鬆弛。

造型： 將麵糰排氣，捲成欖核形(看圖1-8)。

最後發酵：焗盤中間放一個塗油小慕絲圈或布丁模，麵糰五個一組，放入圓模內砌成迴力標形，發酵約30分鐘至兩倍大；除去圓模，噴水，篩上麵粉，在每個小麵糰中間劃一刀，放入已預熱170-180℃的焗爐內，在爐內噴水製造蒸汽效果，焗約30-35分鐘至金黃(看圖9-14)。

小貼士

• 糙米要多煮少許，因糙米飯會黏鍋，煮太少的話會份量不足。

• 在雜貨店買便宜的雙面刀片，穿在長竹籤上便成為實用的麵包剝刀。

• 帆布可在上海街或深水埗的帆布店購買。

1. Knead all ingredients of dough together (except the butter and multi-grain rice). Add pre-ferment dough piece by piece. Knead after each addition until soft and smooth. Add butter and multi-grain rice. Knead until stretchable consistency.

2. Put the dough into a big bowl. Cover with a cling wrap. Let it prove for about 25-30 minutes.

3. Divide the dough into ten equal portions. Flatten each piece with your hands to drive air out. Round them. Set aside to rest for about 20 minutes.

4. Flatten each piece of dough again with you hands to drive the air out. Roll them with your palm into a long oval (see pictures 1-8).

5. Put a greased pudding mould in the centre of a greased round ring mould. Arrange five pieces of long oval dough into a boomerange shape inside the ring mould. Let it prove for 30 minutes until the dough has doubled in size. Remove the ring mould. Spray water and sift some flour all over. Score each piece of small dough at the centre. Bake in a pre-heated oven at 170-180°C for about 30-35 minutes (see pictures 9-14).

Tips

- You may have to cook a bit more multi-grain rice than you need as it tends to stick to bottom of the rice cooker.

- To score the dough easily, you may make your own scalpel – just get a double-sided shaving blade from grocery store. Then insert a bamboo skewer through it.

- You may get canvas from the fabric stores on Shanghai Street or in Shamshuipo.

聖果麵包
Assorted Nut Bread

中種 Pre-ferment Dough	百分比%	克 gram
筋粉 bread flour	70%	358
水 water	40%	205
鮮酵母 fresh yeast	2%	10
海鹽 sea salt	1%	5
脫脂奶粉 skim milk powder	2%	10

酵母溶於水中，再加入筋粉、海鹽、脫脂奶粉搓至柔滑，用保鮮紙包裹麵糰，放入雪櫃發酵17小時。

Dissolve the yeast in water. Add bread flour, sea salt, skim milk powder and knead until soft. Cover the dough in cling wrap. Refrigerate to let it prove for 17 hours.

麵糰 Dough	百分比%	克 gram
筋粉 bread flour	30%	153
脫脂奶粉 skim milk powder	3%	15
合桃 walnuts	15%	77
芝麻粉 ground sesames	5%	26
杏仁角 diced almonds	12%	61
松子仁 pine nuts	15%	77
榛子 hazelnuts	15%	77
海鹽 sea salt	1%	5
砂糖 sugar	10%	51
水 water	28%	143
鮮酵母 fresh yeast	1%	5
無鹽牛油 unsalted butter	8%	41

裝飾	Garnish
杏仁碎適量	flaked almonds

預備
preparation

- 將合桃、杏仁角、松子仁和榛子稍微烘香，待涼，與砂糖一同放入攪拌機內磨成粗粒待用
- 將中種切成小塊
- Bake the nuts briefly. Leave them to cool. Grind them up with sugar in a food processor.
- Cut the pre-ferment dough into small pieces.

A ① ② ③ ④ ⑤ ⑥

B ① ② ③ ④ ⑤ ⑥

C ① ② ③ ④ ⑤ ⑥

D ① ② ③ ④ ⑤ ⑥

E ① ② ③ ④ ⑤ ⑥

重覆A-E步驟圖,直至完成。
Repeat steps A to E until six strands braided into loaf.

做法
method

混合: 將麵糰所有材料(牛油除外)混合,搓揉,逐少加入中種,搓至柔滑,加入牛油再搓至可伸延薄膜。

發酵: 麵糰放入大碗內,蓋上保鮮紙,進行第一次發酵,約25-30分鐘。

分割: 麵糰分割成18等份,排氣,滾圓,蓋上保鮮紙,靜置20分鐘讓麵糰鬆弛。

1. Knead all ingredients of dough together (except the butter). Add pre-ferment dough piece by piece. Knead after each addition until soft and smooth. Add butter. Knead until stretchable consistency.

2. Put the dough into a big bowl. Cover with a cling wrap. Let it prove for about 25-30 minutes.

3. Divide the dough into eighteen equal portions. Flatten the dough with your hands to drive the air out and round each piece of dough. Cover with cling wrap. Set aside to rest for about 20 minutes.

口訣 The pithy formula：

六手辮口訣 The pithy formula for six-strand braiding : 1 over 3, 5 over 1, 6 over 4, 2 over 6

這麵包亦可鬢成三、四或五手辮，口訣如下：

You may also braid the loaf with three, four or five strands:

三手辮口訣 three-strand braiding : 1 over 2, 3 over 2

四手辮口訣 four-strand braiding: 1 over 3, 2 over 3, 4 over 2

五手辮口訣 five-strand braiding: 1 over 3, 2 over 3, 5 over 2

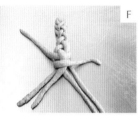   

造型： 將麵糰排氣，搓成長條，中間要略粗，撲上麵粉，每六條一組鬢成六手辮，用雙手修齊頭尾不美的部分，用保鮮紙蓋好(看圖A-L)。

最後發酵： 麵糰發酵約30分鐘至五六成，塗上蛋水，灑上杏仁碎(看圖M)，放入已預熱170-180的焗爐，焗約30分鐘至金黃。

4. Flatten the dough with your hands to drive the air out. Roll the dough with your hands into long strands with tapered ends. Lightly dust them with flour. Braid six strands repeatedly into a loaf. Trim the ends for better presentation. Cover in cling wrap (see pictures A to L).

5. Let it prove for about 30 minutes or until the dough has risen 50-60%. Brush egg wash on top and sprinkle flaked almonds over it (see picture M). Bake in a pre-heated oven at 170-180°C for about 30 minutes.

貼士
tips

- 果仁和糖一起研磨，可避免磨出的油分把果仁黏在一起。
- Grinding the nuts with sugar prevents the nuts from sticking together because of the oil extracted in the grinding process.

果醬、肉醬、濃湯，
再加上天然麥香的麵包，
不就是簡約又美味的一餐嗎？
Jam, meat sauce or
a bowl of soup –
all you need is a healthy dose of
wheat carbohydrate from bread,
for a simple
and delectable meal.

鳳梨果醬
Homemade Pineapple Jam

材料 ingredients

鳳梨(菠蘿)600克	600 g pineapple
砂糖120克	120 g sugar
果膠6克	6 g pectin

做法 method

1. 將果膠和小部分糖調勻。

2. 鳳梨刨成茸，與餘下的砂糖一同煮至水分收乾，煮約半小時，下果膠糖煮至濃稠，期間需不斷攪拌，避免煮焦。

1. Mix the pectin with a little sugar first.
2. Grate the pineapple and keep the juice. Add the remaining sugar and cook until the juice reduces. Add pectin and continue to cook until thick for about 30 minutes. Stir frequently to avoid charring.

貼士 tips

- 使用地捫金鳳梨及台灣香水鳳梨各半，前者取其甜，後者取其香，效果最好。用剩的果醬可塗麵包，亦可伴自製乳酪享用。

- 果膠可令果醬濃稠，在烘焙店有售

- For the best result, use 300 g of Del Monte pineapple and 300 g of Taiwanese Perfume pineapple. The former is sweet while the latter is fragrant. They combine for the best of both worlds. You may use the leftover pineapple jam as a spread on sliced bread or stir it into yoghurt.

- Pectin is a coagulant which is commonly available from baking supply stores.

肉醬
Bolognese

碎牛肉500克	500 g minced beef
洋葱200克	200 g onion
西芹150克	150 g celery
紅蘿蔔200克	200 g carrot
蒜頭15克	15 g garlic
意大利去皮番茄280克	280 g peeled Roma tomatoes (canned)
意大利去皮番茄汁50克	50 g tomato sauce (from the canned tomatoes)
牛油適量	butter
橄欖油適量	olive oil
紅酒300克	300 g red wine
茄膏15克	15 g tomato paste
牛肉清湯500克	500 g beef stock
月桂葉1片	1 bay leaf
百里香適量	thyme
鹽適量	salt

1. 碎牛肉用橄欖油炒至出水,隔去水分。

2. 將蔬菜切成幼粒(意大利去皮番茄除外),用牛油和橄欖油炒香,加碎牛肉再炒香,下紅酒繼續炒至收乾水分。

3. 加入意大利去皮番茄、意大利去皮番茄汁、茄膏、月桂葉、百里香和牛肉清湯慢火煮約兩小時半至肉醬軟身,下鹽調味即成。

1. Pan-fry the minced beef in olive oil. Drain and set aside.

2. Finely chop the vegetables (except peeled Roma tomatoes) and stir-fry them in butter and olive oil. Add minced beef and stir fry until fragrant. Add red wine and cook until it almost dries out.

3. Add peeled Roma tomatoes, tomato sauce, tomato paste, bay leaf, thyme and beef stock. Simmer for 2.5 hours until the meat is tender. Season with salt. Serve.

白汁蘑菇
Creamed Mushrooms

材料
ingredients

蘑菇200克	200 g button mushrooms
乾葱15克	15 g shallot
蒜頭1瓣	1 clove garlic
香草適量	mixed dried herbs
鹽、胡椒適量	salt and pepper
橄欖油適量	olive oil
淡忌廉50克	50 g whipping cream
法式忌廉15克	15 g crème fraîche
拔蘭地5毫升	15 ml brandy
蛋(打勻)1/2個	1/2 egg (beaten)

做法
method

預備

- 將蘑菇、乾葱、蒜頭切碎

- 用橄欖油將蘑菇炒香，下乾葱、蒜頭炒至軟身，下拔蘭地、香草、淡忌廉、法式忌廉，煮至水分收乾，關火，下蛋液煮數秒至汁液濃稠。

貼士

- 法式忌廉通常用以製作醬料，原因是它加熱後不會分離。在製作過程，最重要是要看準時機加入法式忌廉，要一點點的加入，並不時攪拌。

Preparation

- Chop mushrooms, shallot and garlic finely

- Sauté mushrooms in olive oil. Add shallot, garlic and stir fry until fragrant. Add brandy, herbs, whipping cream and crème fraîche. Cook until the sauce almost goes dry. Turn off the heat and stir in the beaten egg. Stir for a few seconds until the sauce thickens further.

Tips

- Crème fraîche is usually used when making cream-based sauces. It doesn't separate after cooking. It's important to add crème fraîche at the right time. Add little by little while stirring continuously to blend it in.

南瓜湯
Pumpkin Soup

材料
ingredients

魚或雞清湯400克	400 g fish or chicken stock
洋葱60克	60 g onion
去皮南瓜500克	500 g peeled pumpkin
去皮蘋果60克	60 g peeled apple
香草適量	mixed dried herbs
橄欖油適量	olive oil
鹽、胡椒適量	salt and pepper
檸檬汁適量	lemon juice
蒜頭1瓣	1 clove garlic
薑1片	1 slice ginger

做法
method

預備
- 南瓜、蘋果、洋葱、薑切粒

1. 用橄欖油先將薑、蒜爆香，下其餘蔬菜炒香。
2. 下清湯、香草，煮至蔬菜變軟約30分鐘，關火。用攪拌機將湯料攪成茸，下鹽、胡椒、檸檬汁調味。

備註
- 份量：4-5碗

Preparation

- Dice pumpkin, apple, onion and ginger finely.

1. Sauté the garlic and ginger in olive oil until fragrant. Add the remaining vegetables. Stir fry until tender.
2. Pour in the stock and add herbs. Bring to the boil and turn the heat down. Simmer gently for about 30 minutes. Remove from heat. Puree the soup in a blender. Season with salt, pepper and lemon juice.

Note

- makes 4-5 servings

雜菌湯
Assorted Mushroom Soup

做法 *method*

預備

- 雜菌、洋葱、乾葱、蒜頭切碎

1. 用牛油將洋葱、乾葱、蒜頭炒香,下雜菌,撒上香草。將雜菌炒至軟身。
2. 注入清湯,用慢火煮約20分鐘,下拔蘭地、淡忌廉、奶再煮至微滾,關火,用攪拌器將湯料打成茸

備註

- 份量:6-8碗

材料 *ingredients*

魚或雞清湯1公升	1 litre chicken or fish stock
雜菌500克	500 g mixed mushrooms
洋葱90克	90 g onion
乾葱30克	30 g shallot
蒜頭15克	15 g garlic
香草適量	mixed dried herbs
鹽、胡椒適量	salt and pepper
牛油60克	60 g butter
淡忌廉70克	70 g whipping cream
奶100克	100 g milk
拔蘭地5克	5 g brandy

Preparation

- Chop mixed mushrooms, onion, shallot and garlic.

1. Sauté the onion, shallot and garlic in butter until fragrant. Then add mushrooms, mixed dried herbs and continue to sauté until the mushrooms are soft.
2. Add stock and simmer for about 20 minutes. Add brandy, whipping cream and milk. Bring soup to a gentle boil. Remove from heat and puree the soup in a blender until smooth.

Note

- makes 6-8 servings

1. **麵粉的牌子重要嗎？**

   用甚麼牌子的麵粉是很重要的，不同牌子麵粉所含的水分、蛋白質和灰質都不同，需要自己拿捏。我用的是日本昭和行的NEON麵包粉，它吸水量比較多，水分可落足。

2. **中種麵糰須要搓出薄膜嗎？**

   中種麵糰的水分佔40%，搓出來的麵糰是很乾的，只要搓至光滑便可放入雪櫃，並不須要有薄膜。

3. **中種麵糰一定要發酵17小時嗎？**

   基本是發酵17小時，讓麵糰能在低溫下充分發展，但不可超過72小時，否則麵糰會變酸。

4. **中種麵糰需要回溫嗎？**

   中種麵糰不用回溫，從雪櫃取出來後，即可切細加入主麵糰內。

5. **怎樣才知道麵糰已攪拌或搓揉足夠呢？**

   要測試麵糰，可以切一小塊麵糰來看看，慢慢拉扯開，觀察拉開時麵糰的情況，如果薄膜拉出時是平滑，斷開時斷口邊緣圓滑就是足夠。相反，如果拉開時麵糰立即斷開，斷口邊緣參差不齊，就要多搓或攪拌一會了。

1. **Does the brand of the flour matter?**

   The brand matters a lot when choosing bread flour.

   Different brands of flour have different contents of water, protein and calcium. You'd have to fine tune the recipes according to each brand for the best results. I personally prefer the Neon brand of bread flour from Japan. It picks up more water than other brands and you may add all the water in the recipe straightaway.

2. **Should pre-ferment dough be kneaded until stretchable?**

   There is only about 40% of water in pre-ferment dough.

   Thus, it tends to be quite dry after kneading. You don't need to work on it until it is stretchable before putting it into the fridge. It's fine as long as it's smooth on the outside.

3. **Does the pre-ferment dough need to be proved under low temperature for exactly 17 hours?**

   You need at least 17 hours for proper proving under such low temperature. You may leave it in the fridge up to 72 hours. The dough proved for longer than 72 hours may taste sour.

4. **Do I need to bring the pre-ferment dough back to room temperature before using?**

   No, you don't need to bring it back to room temperature.

   Just take it straight out of the fridge, cut it into smaller pieces and work it into the main dough.

5. **How do I know if the dough has been mixed or kneaded well enough?**

   To test the dough, cut off a small piece. Then slowly pull it apart with your hands. Observe closely the way it behaves. If the dough looks smooth all the way when you pull it and the torn ends are smooth when it breaks, it has been kneaded well enough. On the other hand, if the dough breaks readily when you pull it, and the torn ends are rough, it has not be mixed or kneaded enough.

6. 為什麼要最後加入油分呢？

因油會阻礙麵糰的擴展。如先加入油，部份酵母被油包圍，使發酵速度減慢，嚴重影響麵糰的膨脹力。

7. 麵糰需要攪拌或搓揉多少時間呢？為甚麼我搓揉一小時麵糰還不光滑和沒拉出薄膜呢？

基本上攪拌或搓揉15-20分鐘已足夠(當然是得心應手後)，但不要超過30分鐘，否則麵糰就開始發酵了，會出現一些小氣泡，麵糰會越搓越爛。多些練習就會熟能生巧了。

8. 為甚麼麵糰好像很軟，無法挺立，很黏手而且拉開時缺乏彈性？

這是過度攪拌的跡象，這時麵包焗出來體積會較扁，包肉粗糙，缺乏彈性。

9. 發酵時如突然要中途停止怎辦呢？

可以將麵糰放入雪櫃，用低溫度將發酵時間延緩一至兩小時是沒大問題的。

10. 氣溫太低或太高怎辦呢？

麵糰發酵宜在26-30℃，若氣溫太低時，除了用溫水打麵糰外，可用一盤暖水坐着麵糰發酵，或者用養魚用暖管，調至適當溫度，放入水中保持溫度，讓麵糰坐着暖水發酵。氣溫太高時可用冷水打麵糰，開空調調節室溫或用冷水坐麵糰發酵。

11. 用甚麼做手粉呢？

用筋粉，它不會黏手，但不可用過量。

6. Why is oil added in the very last stage?

It's because oil retards the leavening action. Oil particles tend to surround the yeasts and slow down their reproduction, which severely hinders the raising power of the yeasts.

7. How long should the dough be kneaded? Why isn't it smooth and turn stretchable even after 1 hour of kneading?

Basically, you only need to knead the dough for 15 to 20 minutes (after you acquired the skills of course). You should never knead it for longer than 30 minutes as the dough would start to prove afterwards. The bubbles formed in the proving process would make it even harder to knead and the dough would turn mushy and inelastic. As always, practice makes perfect. You should be able to do it easily after practicing for a few times.

8. How come my dough looks very soft but limp? It feels sticky but it lacks elasticity when pulled apart.

These are signs of over-kneading. Such dough makes bread that is not raised sufficiently. The interior of the bread would be tough and lack of elasticity.

9. How can I halt the proving process temporarily?

You may put the dough into a fridge. The low temperature would slow down the proving process and you may leave it there for 1 to 2 hours.

10. What should I do if the room temperature is too low or too high?

The best temperature for proving should be 26˚C to 30˚C. When the room temperature is too low, you may use warm water to knead the dough. Alternatively, put the dough in a mixing bowl and then put the mixing bowl into a warm water bath. Cover it up. You may add a fish tank heater to the water bath to keep the temperature. When the room temperature is too high, you may use cold water to mix the dough. Or, you may turn on the air conditioner or put the dough into a cold water bath.

11. What flour should I use to dust my hands?

Use bread flour. It doesn't stick to your hands, but you shouldn't use too much.

12. 如對奶類敏感，可用其他東西代替奶粉嗎？

如對奶類敏感的話，可用無糖豆漿粉代替。

13. 要怎樣分割麵糰呢？

要盡快切割麵糰，切時亦盡量少用手粉，以免手粉滲入麵糰內，亦不宜將它分得太細碎，以免破壞麵糰。最好先把麵糰切成條狀，再切成小塊，用電子磅量度每一個切割好的麵糰。不要過分擠壓麵糰而把氣體排出，使分割困難。

14. 麵包怎樣才算已經焗熟？

在焗爐內已烘出麵包香，出爐後，立即脫模，用手敲一敲麵包底，如發出很空洞的聲音，表示麵包已熟，可放在網架上待涼。

15. 為何中種的材料內沒有糖？

因為糖會令麵糰發酵速度加快，於低溫中種法不須加糖，讓麵糰慢慢地充分發酵。

16. 模具要塗油嗎？

不銹鋼模具可塗一些牛油，比較易上色，而其他不黏模可在模內噴少許食用油噴劑，不塗也可。

17. 怎樣保存麵包呢？

用保鮮袋包好放入雪櫃，可存放一星期；如放進冷凍櫃；可存放一個月。

12. **For those with lactose intolerance, is there any substitute for milk powder?**

You may use unsweetened soybean milk powder instead of milk powder.

13. **Is there any trick in dividing the dough?**

You should do it as quickly as possible. Try not to dust your hands with too much flour as the flour may get absorbed by the dough. You should not cut the dough too small either as it would destroy the gluten structure. The best way to do it is to cut the dough into long strips first and then small pieces. Weigh each piece with a digital scale. Do not squeeze the dough too much when driving the air out. Otherwise, it would be more difficult to divide.

14. **How do I know if the bread is done in the oven?**

You should be able to smell it when the bread it done. Unmould the bread immediately straight out of the oven. Knock the bottom of the bread with your knuckles. It should sound hollow when it's done. You may then leave it on a wire rack for cooling.

15. **How come there isn't any sugar in the pre-ferment dough?**

The pre-ferment dough is meant to be proved slowly for its unique chewiness and yeasty flavour. That's why it is left at the low temperature. In the same line of thought, sugar is intentionally omitted to slow down the proving process. Adding sugar would speed up the yeast fermentation and the dough would lose its unique qualities.

16. **Do all moulds need to be greased?**

You may grease stainless steel moulds with some butter so that the bread browns more nicely. For other moulds, you may grease them with some pan release spray, but it's not mandatory.

17. **How do you keep the bread fresh?**

You may put it in a zipper bag and keep it in the fridge. It lasts for about a week. If you deep-freeze it in the freezer, it lasts as long as a month.

獨角仙的鳴謝

這書能順利誕生，
全賴以下的各方友好協助。

引薦人

Ms. Rachel YAU

學做包餅以來，一直有一個心願，就是將自己的心得和人分享。出書是我夢寐以求但是遙不可及的心願，於是我在網上平台和人分享，得到了很多回應和認識了很多志同道合的朋友，Rachel是其中一位，後來我還發覺她是我的中學同學，我和她有着一份很難言喻的感情。很感激她的引薦，令我有機會一嘗作者夢，和大家繼續分享食物的色香味。

師友

Ms. Louisa HO 、Mr. Johnny CHAN

兩位是香港烘焙專業協會的會長和副會長，都是我最尊敬，對我影響最深遠的師友，多謝他們的教導，令我接觸和認識更多烘焙上的知識，擴闊我的領域。

藍色大門好友

Ms. Cass TUNG、Mr. CHAN Ming-ho、Ms. Tendy YIM、Ms. Irene CHEUNG、Ms. Minnie KAN、Ms. SIU Yuen-sang、Ms. Anne CHAN、Ms. May MOK

多謝他們騰出寶貴時間，無價友誼來協助我籌備這本書，沒有你們這本書恐怕成不了事。

特別鳴謝

西貢大涌口村佳記花園借出拍攝場地

攝影師，雖然他在拍攝期間發高燒至38℃，但他也不去休息，支撐着身體努力完成拍攝工作，他的專業，值得表揚。

《明茶房》泡出真茶味

唯有巧工手製的茶葉，才能給味覺最高的享受。
《明茶房》本著這個信念，與堅守傳統的茶農，
帶您細嚐上品中國茶。手採手造的茶葉，
經我們精心篩選，加上合適的保護，才送到您手。
一絲不苟，只為給您茶的真趣。

《明茶房》專櫃
金鐘太古廣場
Great
中環置地廣場
ThreeSixty
九龍站圓方
ThreeSixty
香港藝術館
香港歷史博物館
香港會展‧設計廊

《明茶房》中國茶專門店〈品茶、小吃、茶班、茶葉、茶具〉

地址： 香港鰂魚涌船塢里8號
　　　　華廈工業大廈12樓D室 (太古地鐵站A1出口)
營業時間：星期一至六上午10時至下午7時
電郵： customer@mingcha.com.hk
網址： www.mingcha.com.hk
熱線： 25202116

天然麵包香 Natural Breads Made Easy

作者　Author
獨角仙　Kin Chan

策劃/編輯　Project Editor
Catherine Tam

攝影　Photographer
Tzee-Man Production

美術統籌及設計　Art Direction & Design
Amelia Loh

出版者　Publisher
Forms Kitchen
香港鰂魚涌英皇道1065號　Room 1305, Eastern Centre, 1065 King's Road,
東達中心1305室　Quarry Bay, Hong Kong
電話　Tel　2564 7511
傳真　Fax　2565 5539
電郵　Email　info@wanlibk.com
網址　Web Site　http//www.formspub.com
　　　　http//www.facebook.com/formspub

瀏覽網站

會員申請

發行者　Distributor
香港聯合書刊物流有限公司　SUP Publishing Logistics (HK) Ltd.
香港新界大埔汀麗路36號　3/F., C&C Building, 36 Ting Lai Road,
中華商務印刷大廈3字樓　Tai Po, N.T., Hong Kong
電話　Tel　2150 2100
傳真　Fax　2407 3062
電郵　Email　info@suplogistics.com.hk

承印者　Printer
中華商務彩色印刷有限公司　C & C Offset Printing Co. Ltd.

出版日期　Publishing Date
二〇〇九年一月第一次印刷　First print in January 2009
二〇一七年一月第六次印刷　Sixth print in January 2017